Antioxidants in Higher Plants

Edited by

Ruth G. Alscher

Associate Professor
Department of Plant Pathology, Physiology, and Weed Science
Virginia Polytechnic Institute and State University
Blacksburg, Virginia

John L. Hess

Associate Professor
Department of Biochemistry and Nutrition
Virginia Polytechnic Institute and State University
Blacksburg, Virginia

CRC Press is an imprint of the
Taylor & Francis Group, an informa business

First published 1993 by CRC Press
Taylor & Francis Group
6000 Broken Sound Parkway NW, Suite 300
Boca Raton, FL 33487-2742

Reissued 2018 by CRC Press

A Library of Congress record exists under LC control number: 92034106

ISBN 13: 978-1-138-55055-1 (hbk)
ISBN 13: 978-1-138-55763-5 (pbk)
ISBN 13: 978-1-315-14989-9 (ebk)

Visit the Taylor & Francis Web site at http://www.taylorandfrancis.com and the CRC Press Web site at http://www.crcpress.com

INTRODUCTION

Oxidative stress in biological organisms is a consequence of life as it is currently defined on this planet. Our purpose is to consider key metabolites in higher plants that have capacity to react as antioxidants at various levels of absorbed energy and oxidation states of oxygen. We have intentionally not considered key enzymes, superoxide dismutase (SOD), catalase, and general peroxidases. Each of these protein classes has been extensively studied and, with the tools of molecular biology, remarkable progress is being made in our understanding of their responses to oxidative stress. Another topic of related interest, photorespiration, is also not included in this volume. It must be recognized, however, that the cycling of carbon through assimilation of CO_2 to key intermediates and its reoxidation to CO_2 provides a means of dissipating energy while minimizing losses of carbon from the plant. It is understood that constitutive levels of enzymes of this pathway in plants are required as a consequence of the current low CO_2/O_2 ratio. This metabolism defines a functional antioxidant pathway requiring chloroplasts, peroxisomes, and mitochondria to: catalyze the consumption of O_2, generate H_2O_2, dismute the peroxide with the peroxisomal catalase, and cycle carbon. However, the extent to which this pathway is modulated in response to other environmental stressors has received little attention.

The substrates that function as electron donors in more general antioxidant metabolism and molecules that absorb energy directly are the focus of our review and comment. These compounds must serve as key structures upon which to develop comparative chemical analogs that may be used to study antioxidant metabolism in order to understand the function and compartmentalization of the natural antioxidants.

A premise for our interest in these compounds is that the oxidative burden on organisms has increased due to anthropogenic factors among which air pollution is a major component. Both light intensity and ambient energies also change over time and most recently include increases in the ambient UV. Fundamental mechanisms for both oxidative and photochemical damage are understood in terms of free radical chemistry or direct energy dissipation through thermal processes. Responses by plants to extremes in temperature and drought stress correlate with responses typically observed from increased oxidative stress. Hence, a continued emphasis on exploring susceptibility of plants to oxidative and photochemical stress is needed in order to understand competitive relationships in ecosystems and effects on overall productivity.

Oxidative stress may be initiated by the generation of organic free radicals that results in self-propagating autooxidation reactions. These radicals form in response to either direct excitation from incident radiation or secondarily from reactions with oxygen radicals or metastable forms of reduced oxygen, i.e., $OH^{\cdot}$, H_2O_2, singlet O_2, and $O_2^{\cdot -}$. The damage to biological systems resides in the formation of other organic peroxides propagated through typical radical chain reactions.

Protection to organisms may be expressed through various strategies. We have learned much from model systems using synthetic antioxidants and inhibitors of oxidation chemistry. Successful protection may result from: direct quenching of the reactive oxygen species, disruption of the free radical chain reaction, or direct absorption of ambient radiation. In each of these reactions it is essential that the final product of the chemistry or the excitation result in a product which no longer carries the properties of the free radical or a reactive excited state. Conceptually, it is when oxidative stress exceeds the normal capacity of the plant cell to dissipate absorbed energy and/or free radical formation that irreversible damage is initiated. The extent to which specific enzymes have evolved to accomplish this protection directly depends on the properties and abundance of specific substrates.

In the investigation of these systems in higher plants much progress has been made in describing relationships between oxygen, ascorbate, and glutathione. The following chapters summarize current research efforts that provide insight into how plants with differential sensitivities to oxidative stress use these metabolites and associated enzymes. This antioxidant capacity is relegated to the aqueous phase of the plant cell more directly than the membrane and likely occurs within organelles as well as the cytosol. However, it needs to be recognized that reactive, water soluble organic peroxides may be released from membrane oxidations and thus react with thiols or other scavengers in the aqueous phase.

Glutathione is a ubiquitous tripeptide that mediates redox cycling between ascorbate and NADPH. It is essential for maintaining the cellular redox status. Much is known about glutathione function, and the rapid changes observed in glutathione contents under oxidative conditions indicate its importance in antioxidant defense. While our knowledge about the regulation of glutathione biosynthesis and degradation in plants is still limited, much recent research has focused on the effects of glutathione on gene expression and on the molecular biology of the associated enzyme, glutathione reductase.

Ascorbate plays a pivotal and clearly defined role in the scavenging of hydrogen peroxide. In the chloroplast, the ascorbate system is important as a detoxification system and as a regulator of electron flow *in vivo*. The reduction of molecular oxygen to hydrogen peroxide, and, eventually, to water in photosynthesis aids membrane energization, also affording protection through triggering processes in the thylakoid that lower intrinsic quantum yield of Photosystem II, helping to prevent photoinhibitory damage. Precise mechanisms involving ascorbate remain unclear for other biochemical and physiological functions of plants. In addition to its role as a primary antioxidant, it is also an important secondary antioxidant. It maintains the α-tocopherol pool that scavenges radicals in the inner regions of membranes. Ascorbate and α-tocopherol share the unique property of a high capacity for free radical scavenging combined with low reactivities with oxygen.

The carotenoids may function to mediate potential damage from excited states of pigment molecules and singlet oxygen. It is expected that these conversions result in dissipation of energy through loss of thermal energy. Protection from photooxidative damage involves the carotenoids in preventing the formation of reactive singlet oxygen. The carotenoids also function in maintaining the structural integrity of the photosynthetic apparatus.

The conversion of violaxanthin to zeaxanthin is mediated by oxygen uptake and cycling back to violaxanthin requires ascorbate. Substantial evidence now exists to correlate levels of zeaxanthin with the capacity of a plant to minimize damaging effects caused by incident radiation in excess of that required to saturate reductive metabolism. It is zeaxanthin that is involved in a thermal energy dissipation process. The positive correlation of the components of the xanthophyll cycle with growth of plants in high light intensity is consistent with a greater capacity for thermal energy dissipation than occurs in shade-grown plants.

Vitamin E, α-tocopherol, is an exquisite, lipophilic, free radical scavenger which, like the carotenoids, exists within biological membranes. As a free radical trap it directly interrupts free radical chain reactions. As with glutathione, mechanisms exist that allow cycling of the quinone (oxidized) and quinols (reduced) using reduced ascorbate or NADPH directly. For these substrates, α tocopherol should function to trap alkylperoxyl free radicals at the expense of normal reducing equivalents available in the cell.

Finally, various classes of phenylpropanoid derivatives, for example, the flavonoids, lignans, tannins, lignins, also function as antioxidants and many of these compounds contribute exclusively to the browning reactions observed in injured or pathogen-invaded plant tissues. Some classes of these compounds are now used commercially as antioxidants. Like the carotenoids, many of these aromatic compounds serve to minimize the effects of direct photobleaching, particularly that initiated by UV irradiation, and irreversible oxidative damage. Much additional information is required about phenylpropanoids, their turnover, and the enzymology associated with their metabolism. As with the antioxidants discussed in the other chapters, the response of phenylpropanoids to environmental factors will lead to a greater understanding of their contribution to protecting plants against oxidative damage.

It is our intention that the reader become more aware of the varied, yet consistent, antioxidant metabolism available to the higher plant cell. Much remains to be learned about the expression and regulation of this metabolism.

Ruth Grene Alscher
John L. Hess

THE EDITORS

Ruth Grene Alscher's central focus of research is on the ways in which green plant cells respond and adapt to environmental cues. She received her undergraduate degree in Biochemistry from Trinity College, Dublin, Ireland in 1965. Subsequently, she studied effects of light on chloroplast composition at Washington University, St. Louis where she received an MA in Botany in 1968. In Dr. Paul Castelfranco's laboratory at the University of California at Davis, her thesis research (Ph.D. in Plant Physiology, 1972) involved chlorophyll biosynthesis in greening tissue. From 1975 to 1977, she was an NIH postdoctoral fellow in the laboratory of Dr. André Jagendorf at Cornell University where she studied the effect of anoxia on the translational capacity of thylakoid-bound polyribosomes. She was a member of the Environmental Biology Program of Boyce Thompson Institute at Cornell University from 1979 to 1988, and an Adjunct Assistant Professor of Plant Biology at Cornell from 1985 to 1988. Her work during the Ithaca years was centered on the responses of antioxidants to the oxidative stress imposed by air pollutants. She has continued to pursue these questions as a faculty member at Virginia Tech. Together with Drs. J. Hess, C. Cramer, and E. Grabau, she is currently investigating the molecular and metabolic roles of the plastid antioxidant proteins glutathione reductase and superoxide dismutase in the resistance of green plant cells to oxidative stress.

John L. Hess has been interested in factors that impact photosynthesis in higher plants throughout his career. His research has focused on oxidative metabolism in both plants and animals and continues to explore how cells accommodate oxidative stress. Following the completion of the A.B. degree in Chemistry at Franklin and Marshall College, Lancaster, Pennsylvania, he pursued a M.S. in Chemistry at the University of Delaware while studying the use of pectinases to release cells from leaf tissue in order to quantify distributions of plastoquinone and ubiquinone. He studied aspects of glycolate metabolism with Dr. N. E. Tolbert and earned a Ph.D. degree in Biochemistry in 1966. He completed postdoctoral work with Dr. A. A. Benson at Scripp's Institution of Oceanography (1966) and with Drs. J. Berry and O. Bjorkman at the Carnegie Institution of Washington Plant Biology Laboratory (1978), where he was able to pursue work on the association of pigments with membranes and the response of plant metabolism to temperature stress. He has been a faculty member in the Department of Biochemistry and Nutrition, Virginia Polytechnic Institute and State University since 1967. He teaches students at all levels and pursues two research interests. How leaf metabolism responds to the combined effects of temperature stress and air pollutant exposure is studied with his colleagues, Drs. R. G. Alscher and B. I. Chevone. The metabolic response of the rat lens to systemic delivery of oxidants and potential sources of free radicals is studied in collaboration with Dr. G. E. Bunce.

THE CONTRIBUTORS

William W. Adams III
Assistant Professor
Department of Environmental, Population, and Organismic Biology
University of Colorado
Boulder, Colorado

Ruth G. Alscher
Associate Professor of Plant Physiology
Department of Plant Pathology, Physiology, and Weed Science
Virginia Polytechnic Institute and State University
Blacksburg, Virginia

Barbara Demmig-Adams
Assistant Professor
Department of Environmental, Population, and Organismic Biology
University of Colorado
Boulder, Colorado

Christine H. Foyer
Research Director
Laboratory of Metabolism
INRA
Versailles, France

Alfred Hausladen
Research Associate
Department of Biochemistry
Duke University School of Medicine
Durham, North Carolina

John L. Hess
Associate Professor
Department of Biochemistry and Nutrition
Virginia Polytechnic Institute and State University
Blacksburg, Virginia

Norman G. Lewis
Director and Professor
Institute of Biological Chemistry
Washington State University
Pullman, Washington

Kenneth E. Pallett
Head
Molecular Biochemistry and Cellular Biology Department
Rhône-Poulenc Secteur Agro
Lyon, France

Andrew J. Young
Senior Lecturer
School of Biological and Earth Sciences
Liverpool John Moores University
Liverpool, England

TABLE OF CONTENTS

Chapter 1

GLUTATHIONE

Alfred Hausladen and Ruth G. Alscher

TABLE OF CONTENTS

0-8493-6328-4/93/$0.00 + $.50

I. INTRODUCTION

Glutathione is widely distributed in plant cells. It appears to be synthesized in both the chloroplast and the cytosol, and to occur in both subcellular compartments at relatively high levels. Its function as an antioxidant is reviewed here. In concert with ascorbate, glutathione acts by scavenging free radicals and hydrogen peroxide. The reactive intermediates form as a consequence of oxidative stresses caused by extremes of temperature, drought, herbicides, or air pollutants. An integral part of the response of glutathione metabolism to oxidative stress appears to include an increase in the levels of the reduced form of the molecule. Both altered gene expression and modulation of enzyme functioning are involved. The possible role of glutathione as an elicitor of transcriptional events associated with stress responses is discussed.

II. PROPERTIES OF GLUTATHIONE AND GLUTATHIONE-DEPENDENT ENZYMES IN THE ANTIOXIDATIVE SYSTEM

A. GLUTATHIONE AND HOMOGLUTATHIONE

1. Properties

The tripeptide glutathione (γ-Glu-Cys-Gly, GSH) is the major low-molecular weight thiol in most plants.[1] Some legumes contain the homologous peptide homoglutathione (γ-Glu-Cys-Ala, hGSH), either exclusively or in combination with glutathione. Klapheck[2] has recently surveyed the occurrence of GSH and hGSH in 13 legume species. Whereas Vicieae and Trifolieae contained both GSH and hGSH, hGSH predominated in Phaseoleae.

The chemical properties of GSH, including its redox chemistry, have been reviewed in detail by Kosower[3] and Mannervik et al.[4] The antioxidant function of GSH is mediated by the sulfhydryl group of cysteine which, upon oxidation, forms a disulfide bond with a second molecule of GSH to form oxidized glutathione (GSSG). GSH can also form mixed disulfides with proteins or other thiols, such as Coenzyme A. The negative redox potential ($E_0' = -0.34$ V) of GSH allows for efficient reduction of dehydroascorbate or disulfide bonds in proteins. While α-tocopherol and ascorbic acid are generally considered to be the major free radical scavengers in biological systems, GSH is also capable of forming a thiyl radical upon reaction with other radicals.[3] To our knowledge, there are no reports on any differences in redox chemistry between GSH and hGSH.

2. Subcellular and Intraorgan Distribution of Glutathione

The GSH concentration in spinach or pea chloroplasts prepared by aqueous isolation techniques has been determined to be between 1 and 4 m*M*,[5-8] but the cytosol also appears to contain high levels of glutathione.[6,10] Gillham and Dodge[8] reported that 90% of total cellular glutathione exists outside the chloroplast in pea leaf cells. Klapheck et al.,[9] on the other hand, have concluded that chloroplasts isolated by aqueous methods have lost a substantial amount of their complement of glutathione, even those plastid preparations which give high values in the ferricyanide intactness test. Using a nonaqueous isolation method, they obtained a glutathione distribution of 35:65 for chloroplast to cytosol. Smith et al.[10] reported 50 to 60% glutathione in the chloroplast, using nonaqueous isolation methods. No experimentally determined values for *in vivo* cytosolic concentrations of glutathione are currently available. Smith et al.[10] have found glutathione in the nonchloroplastic fraction of barley leaves to be 97 to 98% in the reduced form, whereas in chloroplasts only 66 to 76% was reduced. Higher values of GSH were reported for pea chloroplasts, with 90% of the total glutathione being in the reduced form.[6]

Foliar GSH content per gram tissue decreases with leaf age in pea.[6] This observation should be kept in mind when sampling plant material for assaying GSH, as differences in GSH content between leaves of different age can be up to fivefold. Total GSH content of leaves is higher in the light than in the dark; similar changes in roots indicate a function of GSH as a transport form of photosynthetically reduced sulfur.[1,6] The distribution of hGSH and GSH in legumes varies considerably between leaves and roots. Pea leaves contain very little hGSH; in roots, however, the ratio of hGSH to GSH is approximately 1:1.[2]

3. Assay

Enzymatic and HPLC methods for the determination of GSH have recently been reviewed by Anderson[11] and Fahey.[12] Enzymatic determination in a coupled assay system containing glutathione reductase and dithiobis(2-nitrobenzoic acid) (DTNB) is a convenient, inexpensive method which is highly sensitive and specific. After removal of GSH from extracts by 2-vinylpyridine, the same assay can be used to measure GSSG. It should be cautioned, however, that this assay shows twofold higher reaction rates with hGSH than with GSH, which can lead to erroneous results when investigating hGSH-containing plants.[2] A second widely used method is HPLC separation with fluorometric detection after derivatization of thiols with monobromobimane. The method has the advantage of simultaneously measuring other low-molecular weight thiols, such as cysteine, γ-glutamylcysteine, or cysteinylglycine, and has been adapted for the separation of hGSH from GSH.[2] However, the determination of GSSG by this method is less reliable, since it can only be measured as the difference in GSH content before and after dithiothreitol (DTT) reduction of extracts. GSH can be up to 95% of the total glutathione pool;[6] the reduction of GSSG will, therefore, only slightly increase the value obtained for GSH.

B. DEHYDROASCORBATE REDUCTASE

1. Properties

A major function of GSH in the protection of cells against the toxic effects of free radicals is to keep the free radical scavenger ascorbic acid in its reduced, and hence, active form. Dehydroascorbate reductase (DHAR) uses reduced glutathione as an electron donor for the reduction of dehydroascorbate to ascorbate. Its role in ascorbic acid metabolism has been reviewed by Loewus.[13] The enzyme has been purified from spinach leaves.[14,15] It is highly specific for GSH and inhibited by high concentrations of dehydroascorbate. Thiol groups are involved in catalysis, as SH-reagents deactivate the enzyme.[14] Most of the DHAR activity in peas is localized in the chloroplast,[8] but the enzyme from *Euglena gracilis* has been shown to be entirely cytosolic.[16]

2. Enzymatic vs. Nonenzymatic Reduction of Dehydroascorbate

Nonenzymatic reduction of dehydroascorbate by GSH is high at pH values greater than 7[5] and predominates over enzymatic reduction at GSH concentrations above 1 m*M*.[17] The stromal pH in illuminated chloroplasts is about 8, and chloroplastic GSH concentrations are between 1 and 5 m*M*.[5] Thus, the physiological importance of DHAR in unstressed chloroplasts is unclear, since the nonenzymatic reaction rate appears to be capable of keeping the ascorbate pool in its reduced form. However, under oxidative stress, a substantial portion of the glutathione pool can be in the oxidized form,[18] and under these conditions, GSH concentrations might be too low to provide for efficient nonenzymatic reduction of dehydroascorbate. Experiments comparing the participation of NAD(P)H-dependent monodehydroascorbate reductase and DHAR in the regeneration of ascorbic acid have shown that GSH is essential in maintaining ascorbic acid in its reduced form, but it was not demonstrated whether the reduction of dehydroascorbate is enzymatic or nonenzymatic.[14,16]

3. Assay

DHAR is usually assayed by following the GSH-dependent production of ascorbate from dehydroascorbate at 265 nm, but at high background absorptions at this wavelength, a coupled assay system containing glutathione reductase and NADPH may be more suitable.[16]

C. GLUTATHIONE REDUCTASE

1. Overview

Glutathione reductase (GR) catalyzes the NADPH dependent reduction of oxidized glutathione. The enzyme has been purified from a variety of plant sources, and its properties and functions have recently been reviewed by Smith et al.[19] Structure and catalytic mechanism of human and *Escherichia coli* GR are known in detail (cf. Schirmer and Krauth-Siegel[20]). The amino acid sequences of human, *E. coli* and *Pseudomonas aeruginosa* GR are highly conserved,[21] and X-ray crystallography of human and *E. coli* GRs showed good structural similarity between the two enzymes.[22] This high degree of homology between GRs from different sources may prove useful for interpreting structure-function relationships once sequences for plant GRs are known.[23]

2. Subunit Composition

There are conflicting reports on subunit composition of plant GR (cf. Smith et al.[19]). Subunit molecular weights vary from 42 to 72 kDa (cf. Anderson et al.[24]). Using pea chloroplasts, Kalt-Torres et al.[25] found two peptides of 41 and 42 kDa when 1600-fold purified GR was excised from activity-stained native gels and subjected to SDS-PAGE. The addition of protease inhibitors to the chloroplast isolation medium had no effect on the apparent molecular weight of the peptides. Since the native molecular weight of the enzyme was 156 kDa, the authors concluded that pea chloroplast GR is a heterotetramer of the $\alpha_2\beta_2$ type. Mahan and Burke[26] used a similar approach of native/SDS-PAGE, and found protein doublets of 65/63 and 34/32 kDa of corn mesophyll chloroplast GR. The peptides were present in about equimolar amounts, and indicated GR to be a tetramer of four dissimilar polypeptides. Connell and Mullet[27] also reported a 32-kDa peptide that copurified with a 60-kDa GR from pea chloroplasts, but from these results, the question of whether the smaller peptide is a GR subunit or a contaminant could not be answered. The results of other studies, however, suggest that whole leaf GR from spinach,[28] pea,[29] Scots pine,[30] and Eastern white pine,[24] as well as GR from *Euglena gracilis*[31] are homodimers.

3. Isozymes

Animal tissues appear to contain only one form of GR (cf. Schirmer et al.[20]). An interesting new finding is the occurrence of GR isozymes in plant tissues. A first indication of different forms of GR in spinach was presented by Guy and Carter,[32] who found several bands on native polyacrylamide gels stained for GR activity. The banding pattern was different between GR from cold-hardened and nonhardened tissue, indicating some environmental or developmental control of the appearance of multiple forms of GR (see also below). Drumm-Herrel et al.[33] separated two forms of GR in mustard by anion exchange chromatography. The two isoforms appeared to be differentially regulated by phytochrome, and based on the sensitivity of one form to photoxidation, the authors inferred that one isoform was cytosolic and one was plastidic. Purified pea leaf GR could be separated into eight isoforms of identical molecular weight by 2-D gel electrophoresis, with pIs ranging from 6.5 to 5.2.[29] Fractionation of organelles showed distinct isoforms to be present in chloroplasts and mitochondria, and recent findings of other authors confirm these results.[34,35] No evidence was found for posttranslational modifications or protein modification during the purification procedure. Peptide mapping of protease digests showed the isoforms to be very similar, but not identical.[29] Anderson et al.[24] have found two GRs in pine needles with subunit molecular weights of 53 to 54 and 57 kDa.

Significant progress in our understanding of plant responses to oxidative stress (Section V, below) can be expected when future studies address the question of differential regulation of GR isozymes. Such a regulation is not unknown; multiple isoforms of phenylalanine ammonia lyase[36] have been shown to be expressed differently in response to different stresses.

4. Light Activation

Recent work by Foyer et al.[37] provided evidence that GR might be a light-activated enzyme, as GR activity in barley leaves changed with irradiance in the same way as did the chloroplast enzyme NADP-malate dehydrogenase, which is known to be light regulated.[38] This finding contradicts earlier findings in which GR activities in isolated intact spinach or pea chloroplasts were unaffected by illumination.[25,28]

5. Subcellular Localization

Subcellular fractionation studies have shown GR to occur in chloroplasts, mitochondria, and cytosol. Only a minor fraction of the total leaf GR activity is associated with mitochondria,[29] whereas about 80% is located in chloroplasts,[6,8,29] and 20% in the cytosol.[29] Plastids isolated from pea roots, however, contained only 30% of the total GR activity.[39]

6. Assay

GR is usually assayed by following the GSSG-dependent oxidation of NADPH at 340 nm. Smith et al.[40] have introduced an assay in which the reduction of 5,5′-dithiobis(2-nitrobenzoic acid) (DTNB) to thionitrobenzoate (TNB) by GSH is followed at 412 nm. At this wavelength, absorption of compounds present in crude plant extracts is less severe. Furthermore, the assay is more sensitive, as the molar extinction of TNB is greater than that of NADPH.

Some authors have found GR to be inhibited by NADPH.[26,28,32,41] It is likely that this NADPH-dependent loss of activity occurs as the enzyme active site becomes reduced, as has been found for *E. coli* GR.[42] While a physiological role of this NADPH inhibition is unknown, assays should generally be started by addition of NADPH or enzyme to avoid any preincubation of GR with NADPH.

D. OTHER GLUTATHIONE-DEPENDENT ENZYMES

Glutathione S-transferases function in detoxification of xenobiotics by conjugation with glutathione, and have been reviewed in detail by Lamoureux and Rusness.[62] Phytochelatins, the heavy metal binding peptides of plants, are synthesized from glutathione by **phytochelatin synthetase.**[53] In animals, the selenium enzyme **glutathione peroxidase** is responsible for reducing, and thus detoxifying H_2O_2,[63] but despite intensive search, this enzyme could not thus far be detected in higher plants.[64,65] Glutathione peroxidase has, however, been found in the moss *Tortula ruralis*[66] and after selenium induction in the green alga *Chlamydomonas reinhardtii*.[67] A selenium-independent glutathione peroxidase has been isolated from *Euglena gracilis*.[68] Without giving details, Kuroda et al.[69] reported the occurrence of glutathione peroxidase in apple callus cultures.

III. BIOSYNTHESIS AND METABOLISM OF GLUTATHIONE

A. THE γ-GLUTAMYL CYCLE (FIGURE 1)

Glutathione undergoes continuous synthesis and degradation in a cyclic process, the γ-glutamyl cycle. This pathway is well characterized in animals,[43] and recent studies in plant systems have confirmed many of the results found in animals.

Glutathione is synthesized from its constituent amino acids in an ATP requiring a two-step process, and the degradation of GSH leads to the regeneration of its constituent amino acids. In both animals and plants, γ-glutamylcysteine is formed from glutamate and cysteine by the enzyme γ-glutamylcysteine synthetase,[44] and subsequent addition of glycine by glutathione synthetase[9,45-47] leads to the formation of GSH (cf. Rennenberg[1]). In legumes, an alanine-specific homoglutathione synthetase catalyzes the formation of homoglutathione.[47,48]

In animals, GSH degradation is catalyzed by γ-glutamyl transpeptidase (Reaction 1), which transfers the glutamyl group to acceptor amino acids, and by subsequent cleavage of cysteinylglycine by dipeptidases (Reaction 2), and of the γ-glutamyl amino acids by γ-glutamylcyclotransferase (Reaction 3):[43]

$$\gamma\text{-Glu-Cys-Gly} + \text{aa} \longrightarrow \gamma\text{-Glu-aa} + \text{Cys-Gly} \quad (1)$$

$$\text{Cys-Gly} \longrightarrow \text{Cys} + \text{Gly} \quad (2)$$

$$\gamma\text{-Glu-aa} \longrightarrow \text{5-oxoproline} + \text{aa} \quad (3)$$

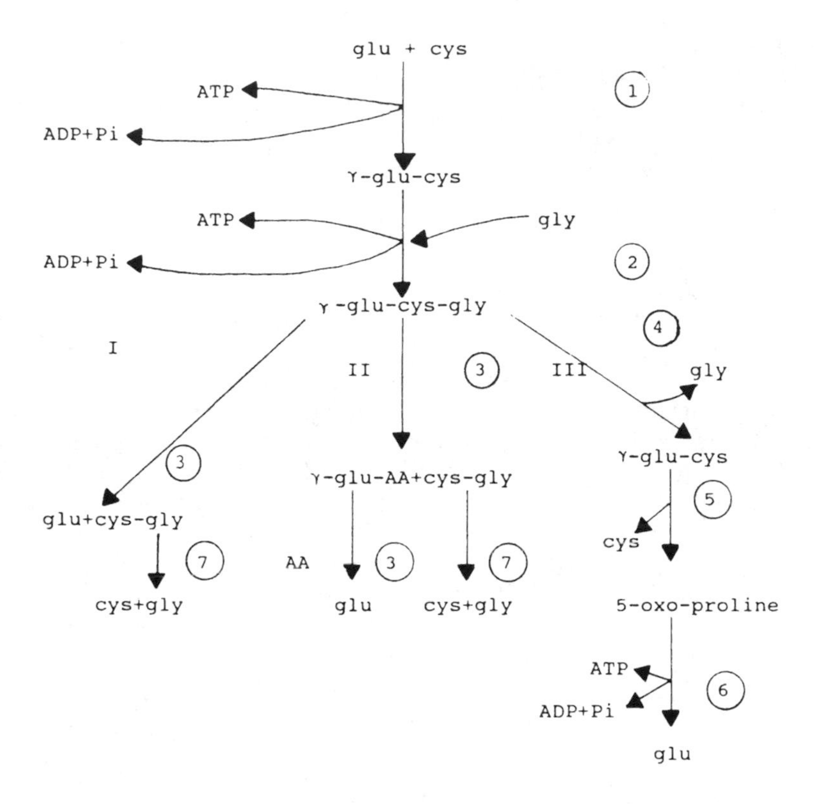

FIGURE 1. Synthesis and degradation of glutathione. Possible path of glutathione metabolism in plant cells. (1) γ-glutamylcysteine synthetase; (2) glutathione synthetase; (3,4) y-glutamyltranspeptidase; (5) γ-glutamylcyclotransferase; (6) 5-oxo-prolinase; (7) dipeptidase. After Rennenberg (1987).

While the biosynthetic pathway seems to be common in plants, animals, and procaryotes, recent studies in tobacco suspension cultures suggest a different pathway for GSH degradation in plant systems (cf. Rennenberg[49]). Although γ-glutamyltranspeptidase has been shown to occur in tobacco suspension cultures,[50] based on labeling studies, a GSH-specific carboxypeptidase (Reaction 4) has been proposed to catalyze the initial step in GSH breakdown, followed by cleavage of γ-glutamylcysteine by γ-glutamylcyclotransferase (Reaction 5):

$$\gamma\text{-Glu-Cys-Gly} \longrightarrow \gamma\text{-Glu-Cys} + \text{Gly} \tag{4}$$

$$\gamma\text{-Glu-Cys} \longrightarrow \text{5-oxoproline} + \text{Cys} \tag{5}$$

The last step in the breakdown of GSH is the cleavage of 5-oxoproline to glutamate by 5-oxoprolinase. The enzyme has been characterized in tobacco and partially purified from wheat germ. In tobacco, 5-oxoprolinase appears to be regulated by sulfur nutrition.[49,51]

In contrast to the animal pathway, γ-glutamylcysteine is a common intermediate in the synthesis and breakdown of GSH in plants. γ-glutamylcysteine might, therefore, constitute a point for regulating GSH content of cells, depending on the relative activities of glutathione synthetase and γ-glutamylcyclotransferase.

B. PROPERTIES AND LOCALIZATION OF ENZYMES OF THE γ-GLUTAMYL CYCLE

γ-GCS and GS have a pH optimum between 8 and 9 and, as ATP-dependent enzymes, exhibit an absolute requirement for Mg^{2+}. For both pea and spinach, 60 to 70% of the γ-glutamylcysteine synthetase activity[44] and 50 to 70% of the glutathione synthetase activity[9,44] have been found in the chloroplast. Glutathione synthetase in tobacco suspension cultures was found to be 24% plastidic.[45] Klapheck et al.[48] attributed only 17% of the homoglutathione synthetase activity in *Phaseolus coccineus* to the chloroplast, based on glyceraldehyde-3-phosphate dehydrogenase activity (a chloroplast enzyme), α-mannosidase activity (a vacuolar enzyme, used as a marker for "cytosol"), and on chloroplast intactness as determined by the ferricyanide method. Considering the difficulties with obtaining intact chloroplasts from Phaseolus, this result should, however, be viewed with caution.

Whereas synthesis of GSH takes place both in the chloroplast and the extraplastidic compartment, the GSH degrading enzymes γ-glutamylcyclotransferase[52] and 5-oxoprolinase[51] in tobacco suspension cultures are entirely cytosolic, based on the distribution of marker enzymes. However, this conflicts with the reported pH optima for these enzymes, which are between 8 and 10.5 (Table 1). The pH in the cytosol is about 7, and whereas γ-glutamylcyclotransferase retains some activity at this pH, 5-oxoprolinase is entirely inactive below pH 8.

The K_M values for the enzymes of the γ-glutamyl cycle are within reported values of intracellular substrate concentrations,[44] and are summarized in Table 2. Considering the high specificity of glutathione synthetase and homoglutathione synthetase for their respective substrates glycine and β-alanine, it is likely that legumes containing both GSH and hGSH[2] also contain both enzymes for their biosynthesis.

C. MODIFICATION OF GSH CONCENTRATION

Experimentally increased or decreased levels of glutathione have been used extensively in animal systems to assess the role of GSH in cellular metabolism (cf. Meister[43]). Buthionine sulfoximine (BSO) and methionine sulfoximine are powerful inhibitors of plant γ-glutamylcysteine synthetase[44] and lead to a depletion of cellular GSH levels.[53,54] L-2-Oxothiazolidine-4-carboxylate (OTC) is an analog of 5-oxoproline which is converted into cysteine by 5-oxoprolinase.[43] Since the supply of cysteine is considered rate-limiting in GSH biosynthesis[1] (see Section III.D, below), OTC treatment leads to a rapid increase in GSH contents.[17,55,56] Glutathione reductase is inhibited by 1,3-*bis*-(2-chloroethyl)-1-nitrosourea (BCNU),[57] but so far no applications of this compound have been reported for plant systems.

TABLE 1
Properties of Enzymes of GSH Metabolism

Enzyme	Source	Mol wt	pH	Ions	Inhibitors	K_M (m*M*)	Ref.
g-GCS	Pea	60,000 2 subunits 34,000	8	Mg^{2+}	BSO GSH	Glu 10.4 Cys 0.19 α-AB 6.4	44
GS	Pea	85,000	8.4		ADP	Gly 0.19 Ala 14 g-GC 0.35 ATP 0.25	47
	Tobacco	—	8–9	Mg^{2+} K^+		Gly 0.31 Ala n.a. g-Gc 0.022	44
	Spinach	—	8.5–9	Mg^{2+} K^+	ADP	g-Glu-AB 0.091 ATP 0.17	46
hGS	*Phaseolus coccineus*	—	8–9	Mg^{2+} K^+		g-GC 0.073 Ala 1.34 Gly 98	48
	Mung bean	—	8.4		ADP	g-GC 0.75 Ala 0.33 Gly 7.5 g-AB 8.3 ATP 0.75	47
g-Glutamyl-cyclotrans.	Tobacco	—	8.7			g-Glu-Met 2.2	52
g-Glutamyl-transpept.	Tobacco	—	8–8.5		g-Glu-analogs	g-Glu-ni-troanalide 0.6	50
5-Oxoproli-nase	Tobacco	—	9.5–10.5	NH_4^+ Mg^{2+} Mn^{2+}		5-OP 0.03	51

D. REGULATION OF GSH BIOSYNTHESIS

Increased *de novo* synthesis of GSH has been documented after a variety of stresses that are related to GSH metabolism[58] (see below). However, little is known concerning changes in enzyme activities of GSH synthesis in response to stress. The properties of partially purified tobacco γ-glutamylcysteine synthetase indicate that this enzyme is regulated by feedback inhibition in the same way as is the animal enzyme. GSH concentrations known to occur *in vivo* completely inhibit the enzyme.[44]

Rüegsegger et al.[53] found increased activities of glutathione synthetase when phytochelatin synthesis was stimulated by cadmium treatment or when glutathione synthesis was inhibited by treatment with the γ-glutamylcysteine synthetase inhibitor BSO. Since Cd^{2+} treatment and BSO both lead to a depletion of cellular GSH, the authors concluded that glutathione synthetase activity is regulated by GSH content. As GSH does not inhibit glutathione synthetase *in vitro,* mechanisms other than feedback inhibition appear to be involved.

TABLE 2
Differential Resistance Linked to Scavenging Enzymes

Species	Stress	Cross-resistance	Enzymes(s)	Location	Ref.
Constitutive *Conyza bonariensis*	Paraquat	Acifluorfen SO_2	SOD, AP, GR	Stroma	128
Conyza bonariensis	Paraquat	High light	SOD, AP, GR	Stroma	158
Lolium perenne	Paraquat	SO_2	SOD, GR	Leaf	159
Induced *Pinus* sp.	SO_2	—	SOD	Needle	160
Phaseolus sp.	Acifluorfen	—	GR	Leaf	81
Hordeum sp.	Drought	—	GR, AP	Leaf	122
Spinacea sp.	Ozone	—	AP	Leaf	78
Spinacea sp.	Cold	Light	SOD, DHA, MDA	Stroma	123

Thus, both enzymes involved in GSH biosynthesis seem to be activated by low GSH content, although the mechanisms for this activation are different.

GSH degrading enzymes in yeast respond to changes in sulfur, nitrogen, or amino acid nutrition (cf. Rennenberg[49]), but to date there are no reports on activities of these enzymes under stress conditions when GSH levels are known to be affected.

E. EFFECT OF LIGHT ON GSH BIOSYNTHESIS

The conditions required for optimum activity of γ-GCS and GS, high Mg^{2+} concentrations, and pH values around 8, are known to occur in illuminated, but not in darkened chloroplasts.[46] Several studies have shown that increases in GSH contents are indeed light dependent, but the mechanisms of this light dependence are unclear. Shigeoka et al.[16] provided evidence that light-dependent GSH synthesis in *Euglena gracilis* requires *de novo* nuclear protein synthesis, since it could be prevented by cycloheximide. In spruce, GSH content undergoes diurnal fluctuations with higher levels during the day and lower levels during the night.[59] When branches were enclosed in a black bag, the increase in GSH during daytime could be prevented or reversed. One possible explanation of these results would be a decreased activity of GSH synthesizing enzymes in the dark as Mg^{2+} concentrations and pH decrease, but as the authors point out, diurnal variations in the export of GSH out of the needles may also be responsible for the observed changes.[59] The increase in GSH might also be a secondary effect of light, since photoreduction of O_2 during photosynthetic electron transport produces O_2^- and H_2O_2 in the chloroplast[60] (see below). Increased production of H_2O_2[18] or other oxidants are known to increase GSH content[58] (see below).

Evidence for GSH synthesis in the dark comes from experiments by Buwalda et al.,[61] who found increased GSH content in spinach fumigated with H_2S in the dark. They also observed that low levels of γ-glutamylcysteine accumulated in the dark, which were converted into GSH upon exposure to light. When either glycine, glycolate, or glyoxylate were supplied to darkened spinach leaves, the accumulation of γ-glutamylcysteine was prevented, and the GSH content increased.[54] The authors concluded that the supply of glycine from photorespiration is involved in limiting GSH biosynthesis in the dark. However, H_2S-induced GSH biosynthesis was markedly higher in the light than in the dark.

In conclusion, different mechanisms appear to limit GSH biosynthesis in the light and in the dark. While properties of γ-GCS and GS suggest that at least the chloroplastic fraction of these enzymes might be inactive in the dark, *in vivo* experiments show that the substrate for GS, glycine, limits GSH biosynthesis in the dark. In the light, when the supply of glycine by photorespiration is adequate, the γ-GCS substrate cysteine appears to be rate limiting, since both H_2S fumigation[54,61] and OTC treatment[17,55,56] increase cysteine content and lead to an accumulation of GSH.

IV. ACTIVE OXYGEN IN PLANTS

A. FORMATION OF ACTIVE OXYGEN

1. The Role of Oxygen in the Formation of Free Radicals

The oxygen molecule in its ground (triplet) state is relatively unreactive, since, although it does possess two unpaired electrons, they are in parallel spins. A pair of electrons could not be added without a spin inversion. For spin reversal to take place, energy input is necessary. Within the illuminated chloroplast, this can take place through the agency of a photosensitizer such as chlorophyll, giving rise to singlet oxygen. If either spin reversal or electron donation/reduction (the more likely event) takes place, the resultant molecules become much more reactive.[70]

Complete reduction of oxygen to water is a four-electron process, which occurs sequentially. Thus, one, two, or three electron-reduced intermediates are possible. The one electron intermediate is superoxide O_2^-. It can react with the two electron intermediate, hydrogen peroxide, to produce ·OH, a highly reactive molecular species in the Haber-Weiss reaction:

$$O_2^- + H_2O_2 \longrightarrow \cdot OH + OH^- + O_2$$

This reaction is thermodynamically slow. However, in the presence of an oxidation-reduction catalyst, such as iron, significant production of ·OH can occur. The iron-catalyzed reaction is the Fenton reaction:

$$O_2^- + Fe^{2+}\text{-chelate} \longrightarrow O_2 + Fe^{2+}\text{-chelate}$$

$$Fe^{2+}\text{-chelate} + H_2O_2 \longrightarrow \cdot OH + OH^- + Fe^{3+}\text{-chelate}$$

Oxygen is most commonly reduced via a multi-step univalent pathway. It is these intermediate oxyradical species which are highly reactive,[71] since they can react with proteins, nucleic acids, and/or lipids, potentially causing denaturation, mutagenesis, and/or lipid peroxidation. Singlet oxygen can also cause lipid peroxidation. Given the constant presence of oxygen in their environment, the possibility of such destructive oxidative reaction exists at all times in aerobic cells.

2. The Mehler Reaction

A first product of the univalent reduction of oxygen is the superoxide anion (O_2^-). Donation of a second electron brings about the formation of hydrogen peroxide. Both of these molecular species are known to be produced by illuminated chloroplasts. Superoxide is formed in a reaction, named after Mehler who first described it in 1951, in which molecular oxygen is reduced by ferredoxin on the reducing side of Photosystem I (original Mehler reference). This reaction occurs in all illuminated, functionally active chloroplasts, but under circumstances where availability of $NADP^+$ is limiting; for example, at low CO_2 levels, the Mehler reaction will occur to a greater extent.[72]

3. The Action of Xenobiotics Involves Oxyradical Production

The production of reactive oxygen species is an important part of the mode of action of environmental stresses such as the pollutants ozone[74,76,78] and sulfur dioxide,[73,74,77] as well as the photoactive compounds that directly absorb light and generate radicals (e.g., cercosporin).[75] The bipyridilium compounds,[86,87] such as paraquat, which divert electrons from ferredoxin (photosystem I), and the diphenyl ethers,[81,84] which cause a build-up of protoporphyrin IX, all cause a build-up of oxidants in photosynthetic tissue. Each case is different with respect to the individual chemistry involved. For paraquat and diquat, both dicationic bipyridilium herbicides with redox potentials close to that of ferredoxin, the molecule is reduced to a radical through reduction by Photosystem I in competition with ferredoxin under conditions where linear photosynthetic electron transport is taking place. Monocationic paraquat is rapidly oxidized by molecular oxygen, forming superoxide anion and regenerating dicationic paraquat.[70,85] Paraquat radical *per se* can directly reduce H_2O_2 in a reaction that is analogous to the Fenton reaction, thereby leading to the production of $\cdot OH$.[86]

$$PQ^+ + H_2O_2 \longrightarrow PQ^{++} + OH\cdot + OH^-$$

Babbs et al.[87] reported that a normal herbicidal application of paraquat for weed control produces an amount of $\cdot OH$ equivalent to 33,000 rads of γ irradiation, i.e., an amount approximately three times that necessary to kill 97% of cells grown in tissue culture.

The oxidizing air pollutant ozone forms oxyradicals and singlet oxygen in aqueous solution, and comparable reactions have been demonstrated to occur in ozone-exposed plant tissue.[88-90] In the case of sulfur dioxide, Asada and Kiso[91] showed that exposed illuminated chloroplasts formed oxyradicals at a rate comparable to a chain reaction initiated by a reaction between sulfite and superoxide.

Aerobic organisms constantly cope with the potential of attack by oxyradicals. The toxicity of oxidizing molecular species of exogenous origin is likely caused by overriding existing resistance mechanisms. Only when the protective capacities of these mechanisms are overwhelmed should injury occur.

B. SCAVENGING OF ACTIVE OXYGEN IN THE CHLOROPLAST

Superoxide anion is removed through the action of the superoxide dismutases, which results in the production of hydrogen peroxide. Photosynthetic plant cells contain at least five distinct forms of this enzyme, which fall into two classes, Cu, Zn- and Mn-containing forms.[92,93] Hydrogen peroxide, the product of superoxide dismutase (SOD) activity, is itself toxic. It can inactivate –SH-containing regulatory enzymes[94] or react with superoxide to form the much more toxic hydroxyl radical. It is clear that further metabolism of the molecule must take place if this damage is to be avoided. Since superoxide arises within the chloroplast as a result of pseudocyclic electron transport, even in the absence of exogenous oxidative stress, it is not surprising to find that SODs are plentiful and active there.

Hydrogen peroxide, produced as a result of SOD activity, is removed through the action of a metabolic cycle located in the stroma involving successive oxidations and rereductions of ascorbic acid (AsA), glutathione (GSH/GSSG), and NADPH.[14,95-97] Rapid operation of the cycle and regeneration of its intermediates through photosynthetically generated reductant is necessary, since at least one of its constituent enzymes, ascorbate peroxidase, is inhibited by hydrogen peroxide itself.[98] Each step has been characterized enzymatically. The enzymes involved, besides SOD, are ascorbate peroxidase (AP),[37] NAD(P)H-dependent monodehydroascorbate reductase,[14] dehydroascorbate reductase, and glutathione reductase (GR).[14,83,97] For each enzyme of the scavenging cycle which has been studied, its kinetics have been shown to be such as to ensure its activity under conditions known to exist *in vivo* in photosynthetically functional chloroplasts. In the case of chloroplast glutathione reductase, its Km for oxidized glutathione was found to be 11 μM, and that for NADPH 1.7 μM, while NADPH concentrations *in vivo* in the stroma are about 1.7 mM in the light and 0.7 mM in the dark.[25] Thus the enzyme will be quite active under both environmental conditions. In fact, it has been shown that glutathione exists *in vivo* largely in the reduced state in both the dark and the light.[6] The other components of the chloroplast photoscavenging cycle are present at very high concentrations in the stroma[83] (see Section II, above).

C. THE CYTOSOL

Detailed information is available for the scavenging mechanisms of the chloroplast because of its relative tractibility as an object of study. Information is available concerning plastid volume, and also concerning concentrations of metabolites existing in the light and the dark in many cases. It is possible to obtain intact, uncontaminated preparations for the study of metabolic pathways, such as the one associated with free-radical scavenging. The picture is not so clear in the case of the rest of the plant cell.

The entire glutathione-ascorbate cycle may exist in the cytosol, as well as in the chloroplast,[6] and in nonphotosynthetic tissue, such as roots,[39,99,100] although the amount of enzyme activity in these locations appears to be lower than in the chloroplast. Most importantly, a demonstration of enzymatically catalyzed scavenging of hydrogen peroxide by components of any subcellular compartment other than the chloroplast or peroxisome is not available. Glutathione reductase and SOD have both been reported to occur in heterotrophic tissue, as well as in the mitochondrion and the cytosol.[101,102]

Glutathione and ascorbate have both been shown to have the potential for an antioxidant role in conjunction with α-tocopherol. Since α-tocopherol is the major free radical scavenger of thylakoid membranes, this finding is of particular relevance for chloroplast metabolism. The direct interaction of α-tocopherol and ascorbate is well established.[103] An interactive role for glutathione with membrane constituents was established by Barclay[104] in an artificial liposome membrane system where it was shown that glutathione acts to spare the lipid-soluble molecule by lowering the rate of chain initiation of free radical formation. This occurred when peroxidation was initiated in the aqueous phase.

Wise and Naylor[30] demonstrated the protective role of antioxidants by comparing the response of glutathione and ascorbate to cold stress in a chilling-sensitive species (cucumber) and a chilling-resistant species (pea). In pea, little photooxidation occurred as a consequence of chilling stress, GSSG did not accumulate, total glutathione decreased only slightly, and lipid peroxidation did not take place over a 12-h time period in the light. In cucumber, on the other hand, chilling resulted both in extensive photooxidation and lipid peroxidation over the same time period, and in substantial accumulation of GSSG and decreases in total glutathione.

V. RESPONSES AND ADAPTATIONS TO OXIDATIVE STRESS

A. OXIDATIVE STRESS LEADS TO SPECIFIC GENE EXPRESSION IN BACTERIA

Exposure of prokaryotes to environmental stress is known to evoke global cellular responses that protect and help adapt to oxidative stress.[105] This response requires the induction of several proteins whose expression is controlled at the gene level. Two oxidative stress responsive regulatory loci have been identified in *E. coli*. These are H_2O_2-inducible *oxyR* and superoxide-inducible *soxR*. The protein OxyR encoded by *oxyR*, under oxidizing vs. reducing conditions, undergoes a conformational change that alters its interaction with the promoter regions of these genes.[105] Parallel systems of oxidative stress regulation may exist in plants.

Schmidt and Kunert[106] compared the response of two strains of *E. coli*, differing with respect to their complement of glutathione reductase, to an agent of lipid peroxidation. The presence of the enzyme clearly conferred resistance to oxidative stress; in fact, if a plasmid containing the gene for GOR was introduced into the GOR^- strain, it acquired resistance, and the enzyme was induced up to fivefold over the control value.

Resistance has been correlated with higher "constitutive" levels of antioxidants in some higher plant systems. Superimposed upon this regulation are inducible responses specific to the cytosol or plastid. Transcriptional and translational events are implicated in these metabolic responses to oxidative stress. Tanaka et al.[107] demonstrated an increased production of the glutathione reductase on Western blots in response to ozone stress in spinach.

B. COORDINATED CHANGES IN COMPONENTS OF THE SCAVENGING CYCLE

Recent data show elevated levels of SOD, GR, and AP, and of glutathione in plants exposed to oxidizing air pollutants. Results of Western blots indicate an increased synthesis of both chloroplast and cytosolic forms of SOD in the tolerant cultivar.[108]

Other results suggest that changing a single component of the ASC-GSH cycle is not sufficient to reduce the toxicity of superoxide. Transformed tobacco plants carrying a petunia rbcS-SOD-1 chimeric gene and with 50-fold the normal activity of SOD-1 (plastid form) do not exhibit increased resistance to paraquat.[109]

Under conditions of oxidative stress, such as exposure to O_3, SO_2, herbicides, heat shock, or drought, the level of reduced glutathione in foliar tissue has been shown to increase above control levels.[78,110-116] In the case of herbicide treatment, the increase is severalfold and rapid. The increase, which is light dependent in leaves, appears to occur as the result of an increase in glutathione biosynthesis,[18] and has been shown in some experimental instances to confer resistance to oxidative stress.[78,111] Mehlhorn et al.[117] report a twofold increase in levels of ascorbate peroxidase and glutathione reductase in peas which had been exposed for 3 weeks to either ozone alone or to mixtures of sulfur dioxide and ozone. Treatment of three species of aquatic plants with an array of antioxidants, including glutathione, resulted in a delay in the changes accompanying senescence.[118] Diurnal fluctuations in GSH levels in spruce needles and soybean leaves have been observed.[59,119] Corresponding diurnal fluctuations in the susceptibility of foliar tissue to oxidants might be expected to occur as a consequence, and have, in fact, been reported for ozone exposure.[120] In the case of heat shock, Nieto-Sotelo and Ho[115] demonstrated that a net synthesis of GSH occurred from cysteine in heat-shocked maize roots. However, they did not demonstrate a correlation of increased glutathione biosynthesis with resistance to heat stress.

The results of Smith[10,18,121] have provided important information concerning the biosynthesis of glutathione in plants under oxidative conditions. In a catalase-deficient mutant of barley under photorespiring conditions, and in wild-type barley treated with aminotriazole, an inhibitor of catalase, glutathione accumulated to levels three times higher than those occurring in nonphotorespiring plants kept in a CO_2-enriched atmosphere. Glutathione afforded some protection to the photosynthetic apparatus in the mutant under these conditions. A larger fraction of the glutathione produced in the mutant was initially in the form of GSSG. These results suggest that the oxidation of GSH to GSSG may have been the trigger for increased glutathione synthesis. The release of feedback inhibition of the pathway through an initial decrease in GSH as it is oxidized to GSSG may be involved, as is suggested by Smith and his co-workers.

The aminotriazole treatment had the same effect in tobacco and soybean. The increase in total glutathione, which was shown to be a net synthesis from sulfate, occurred only in the light. When soybean shoots were treated with aminotriazole in the light, followed by a 20-h dark incubation, no further increase in total glutathione occurred during the dark period, but the GSSG initially present had been converted to GSH.[10,18] Thus, glutathione reductase acted in the dark to reduce GSSG to GSH, a process which requires NADPH, probably furnished by the pentose phosphate pathway. The apparent cessation in glutathione biosynthesis in the dark may be an expression of the light-dependence of the pathway (see Section II, above). Alternatively, GSH formation may have been due to the cessation of hydrogen peroxide production, and hence the source of oxidative stress, in the dark. Since much of the GSH present in the pea leaf cell is located in the cytosol, the hydrogen peroxide produced in the chloroplast as a result of the action of methyl viologen may have diffused across the chloroplast envelope to oxidize the GSH pool present there.

The herbicide paraquat competes for electrons from the primary electron acceptor of Photosystem I, resulting in enhanced production of superoxide. Differential sensitivity to paraquat might therefore be associated with the stromal scavenging cycle. Two biotypes of *Conyza bonariensis* have been described that show differential resistance to paraquat.[124,125] Intact chloroplasts from the resistant R-biotype had higher levels of AP, GR, and SOD than did plastids obtained from the susceptible S-biotype. Isolated thylakoids from both biotypes were equally susceptible to paraquat.[126] Thus, the site of paraquat resistance must be the stroma, and the evidence suggests that it is associated with enzymes of the ASC-GSH pathway. The R-biotype is also differentially more resistant to sulfur dioxide and transient photo-inhibitory stress.[124,127,128]

Inhibitors of Cu/Zn SODs and AP (e.g., diethyldithiocarbamate [DDC], which chelates copper) reverse the paraquat resistance of the R biotype *Conyza*, clearly indicating the critical role of plastid antioxidant enzymes in resistance.[129] These compounds also increased the paraquat susceptibility of the S-biotype, other plant species, cultured plant cells, and isolated plastids.[130]

C. CROSS-RESISTANCE

Glutathione reductase activities have been positively correlated with cross-resistance of higher plants to oxidative stresses such as paraquat and the diphenylethers,[82] as well as with resistance to heat and drought stress.[131] Exposure of a chilling-sensitive strain of *Chlorella ellipsoidea* to paraquat results in increased chilling resistance and higher SOD activities.[132] A specific form of SOD appears to be involved. Other evidence suggests that AP, SOD, and GR are all involved in oxidative stress resistance, and that one dominant nuclear gene affects resistance by pleiotropically controlling the levels of all three stromal enzymes.[127,133] Genetic analyses of the *Conyza* R- and S-biotypes indicate that the paraquat-resistant phenotype is conferred by a single dominant locus.[134] Elevated activities of plastid GR AP, and SOD co-segregated with resistance in both the F_1 and F_2 generations.

Hence, a common regulatory mechanism (e.g., a positive regulatory protein) appears to mediate the coordinate expression of these enzymes.

It is clear that the stromal scavenging pathway is an important antioxidant defense mechanism in plant cells conferring resistance against oxidative stress regardless of its origin (Table 1). Differences between biotypes with regard to resistance suggest the pleiotropic control of at least three of the enzymes involved. By analogy to bacterial systems, a regulatory gene may coordinate the expression of genes for stromal AP, SOD, and GR.

D. COLD STRESS INVOLVES FREE RADICALS

Several studies have shown that free radicals are produced when plants are exposed to low temperatures in the light.[30,135] In the case of conifers, cold tolerance is acquired during the early fall. This process is both photoperiod- and temperature-dependent. Hardening is accompanied by an alteration in thylakoid lipids both with respect to the content of unsaturated fatty acids and the proportion of mono- and digalactosyl residues associated with the membranes.[80] Glutathione and glutathione reductase accumulate to higher levels in conifer needles during the hardening season[136] and also in leaf tissue of frost-hardened spinach.[137] This may reflect a stress response in the cells of hardening cells, due to the increased rates of oxidative processes occurring there. Alternatively, the increase in glutathione may be controlled as part of the overall process of hardening by some other mechanism.

Anderson et al.,[138] reported substantial increases in AP activities in the fall in Eastern white pine (65-fold, year 1; 40-fold, year 2). SOD activities increased also, but to a lesser degree (4.7-fold, year 1; 1.44-fold, year 2). Hausladen et al.,[139] found a correlation in time between the acquisition of cold tolerance and increases in GR activity in red spruce needles. In order to maintain GSH under winter conditions, GR must be active at ambient temperatures. Guy and Carter[32] showed that partially purified enzyme obtained from hardened spinach plants had a greater affinity for GSSG at 5°C than did enzyme present in unhardened plants. The lower temperature optimum of substrate binding of GR in hardened plants may result from alterations of existing enzymes or from synthesis of a distinct isoform having altered kinetic properties from GR expressed in unhardened plants. Guy and Carter[32] have demonstrated that two isozymes of altered electrophoretic mobility do, in fact, appear in hardened spinach, which are absent in unhardened tissue. These data documenting *de novo* synthesis of GR isozymes on exposure to cold imply the action of an environmental trigger on the pattern of gene expression itself. In no case, however, is a primary sequence of a plant GR known. The transition to the winter-hardened state in red spruce also appears to involve the appearance of a form of GR with altered electrophoretic mobility.[23,139]

Mohapatra et al.[140] have demonstrated that cold adaptation in alfalfa was accompanied by elevated mRNA levels for several genes termed cold-acclimation specific (CAS). Relative expression of these CAS genes, as measured by transcript abundance using cloned CAS cDNA sequences, was correlated with the degree of cold tolerance of several alfalfa cultivars. Because the red spruce and spinach data both suggest the existence of a form of GR specific to cold adaptation, sequences encoding GR may be among the group of CAS genes.

E. THE CENTRAL ROLE OF GLUTATHIONE AND OTHER COMPONENTS OF THE SCAVENGING PATHWAY IN PLANT RESPONSES TO AIR POLLUTANTS AND OTHER OXIDATIVE STRESSES

Elevated levels of SOD and GR have been reported in plants exposed to SO_2, O_3, and high levels of O_2. Antioxidant responses were studied in two cultivars of pea known to be differentially sensitive to SO_2.[141] The results suggest that SO_2 enhances both glutathione biosynthesis, GR, and SOD activity, with the tolerant cultivar showing greater responses. The cultivars also show cross-resistance to paraquat. Total glutathione accumulated more rapidly on exposure to the pollutant in the tolerant cultivar with a corresponding increase in GR activity. GSSG did not accumulate in the leaves of either cultivar. SOD activities increased greatly on exposure in the tolerant cultivar, whereas little or no change was observed in the sensitive cultivar, Nugget. This increase occurred subsequent to the increases in glutathione levels and GR activity. Since an oxidative chain reaction of sulfite involving the increased production of superoxide has been implicated in the damage of plants exposed to SO_2, increased SOD activity may alleviate the damage by scavenging the superoxide ion and terminating this chain reaction. Results obtained using antibodies against spinach SOD (provided by K. Asada, Kyoto University, Japan) indicated an increased synthesis of both chloroplast and cytosolic forms of the enzyme in the tolerant cultivar.[108]

Glutathione reductase activity increased in both resistant and sensitive pea cultivars exposed to SO_2 compared with their respective controls, but the increase was more rapid and greater in the resistant cultivar.[141] Onset of an appreciable increase in GR activity corresponded to an increase in GSH content in the tolerant cultivar, Progress. Our results suggested that SO_2 enhances both glutathione biosynthesis and GR activity, with the tolerant cultivar showing greater responses. Spinach, fumigated with 0.07 ppm ozone, showed significant increases in GR protein levels and GR enzyme activity.[107] Thus, it is likely that the SO_2-induced increases in these enzymes in pea were also due to *de novo* synthesis.

A similar chronological sequence of events has been observed in poplar leaves exposed to O_3.[142] The degree of stress imposed by ozone, as expressed by decreases in apparent photosynthesis, correlated with increases first in levels of glutathione and subsequently in the activity of SOD. The degree of inhibition of photosynthesis appeared to be related to the extent of oxidation of the intracellular glutathione pool (as expressed in the GSH to GSSG ratio). Postexposure, at a time when photosynthesis had recovered to 75% of its original rate, GSH levels and SOD activity remained high, and GSSG had decreased. This suggests that the antioxidant-resistance mechanism responds with a successful defense to the brief O_3 exposure, allowing carbon assimilation to resume when toxic free radicals have been effectively removed. If glutathione, GR, and SOD do provide an adaptive function by scavenging free radicals, as has been suggested, one would expect that plants which had been preexposed to oxidant, with elevated glutathione and SOD, to be less sensitive to higher pollutant exposure. Indeed, poplar leaves exposed to 0.1 ppm SO_2 and with increased activities of SOD were more resistant to a subsequent exposure to 2.0 ppm SO_2 than were control leaves.

The response of SOD activity to SO_2 exposure in pea was slower than that of GR activity and total glutathione content. A causative chain of events involving glutathione biosynthesis/metabolism may be involved in the induction of SOD in the tolerant pea cultivar exposed to SO_2.

Matters and Scandalios[143] have shown an increase in the amounts of cytosolic SOD isozymes, SOD-2, and SOD-4, but not of the chloroplast (SOD-1) or of mitochondrial (SOD-3) isozymes in maize leaves exposed to 90% O_2. This increase in cytosolic isozymes was due to increases in levels of sod-2 and sod-4 polysome-bound mRNA. The increase in SOD activity in maize coincided with the appearance of visible damage and decrease in AP activity, suggesting that toxic concentrations of H_2O_2 had accumulated in the chloroplasts. This increase may be linked to the lack of induction of SOD-1. The accumulated data suggest that SOD plays a significant role in resistance to oxidative stress, and this response may involve changes in gene expression.

A crucial aspect of the scavenging pathway, therefore, is the inducibility of its constituent enzymes and the increased synthesis of at least one of its components on exposure to oxidative stress. The operation of the scavenging pathway must be energy-requiring, and as such, should drain reductant and nucleotide triphosphates away from anabolic metabolism. Exposure of hybrid poplar to subacute levels of ozone was positively correlated with increases in SOD and in total glutathione.[142] These increases did not involve an inhibition of electron transport, since the rate of production of NADPH in the light was unaffected by this exposure regime. Since carbon fixation was inhibited and electron transport remained constant, it may be that the resources of the photosynthetically active cell are being diverted from production to defense. The decreased biomass in ozone-exposed crop plants substantiate this competition at the whole plant level. Only when these defense systems have been overwhelmed should toxic molecular species persist long enough to damage essential processes and macromolecules there, which is supported by earlier studies using intact illuminated spinach chloroplasts.[144,145] Concentrations of ascorbic acid and GSH remained constant in the absence of added CO_2, when the Mehler reaction was expected to proceed at a maximal rate. However, if additional H_2O_2 was added, the ascorbate and glutathione pools became quickly oxidized.[144,145]

Numerous other studies have shown that an increase in GSH levels is an early response of plants to oxidative stress (cf. Alscher[58]); the glutathione molecule and the pathway which controls its synthesis would appear, therefore, to be important parts of a resistance mechanism which responds to oxidative stress.

Recently, the *E. coli* gene for glutathione reductase has been expressed in the cytosol of transgenic tobacco plants, which resulted in an up to tenfold increase in total leaf GR activity.[146,147] Foyer et al.[147] found no difference in the effects of methyl viologen on CO_2-dependent O_2 evolution between controls and transgenic plants, but the ascorbate pool was kept in a more reduced state in transgenic plants under oxidative conditions, indicating that GR may limit the capacity of the ascorbate/glutathione pathway.

Aono et al.[146] reported less visible injury after methyl viologen treatment of transgenic tobacco with a 3.5-fold higher GR activity. They concluded that GR, unlike SOD,[109] can be a limiting factor in resistance to methyl viologen. However, Aono et al.[146] found no increased resistance to ozone in transgenic plants. This may reflect the different modes of action of methyl viologen and ozone, as well as the fact that the elevated GR levels were entirely cytosolic. Foyer et al.[147] also observed increased amounts of the bacterial GR in transgenic tobacco after methyl viologen treatment, when compared to untreated transgenic plants. Since the GR gene was under the control of the noninducible CaMV promoter, the authors suggested that the increased amount of bacterial GR may reflect increased translation of GR-mRNA under oxidative conditions.

VI. OXIDIZED AND REDUCED GLUTATHIONE AS MODULATORS OF PLANT METABOLISM

A. EFFECTS ON ENZYMES

Some time ago, it was proposed that GSSG might act to inactivate light-modulated stromal enzymes, such as fructose 1,6-bisphosphatase, through an oxidation of their –SH groups.[148] However, subsequently, it was shown by Halliwell and Foyer[28] that, in fact, *in vivo* in-the-dark GSSG levels never rose to a point where such an oxidation could occur. Some interesting new results suggest that the reverse is possible, and that GSH may act to activate at least one light-modulated enzyme.[149] Concentrations of GSH in the 3 to 10 m*M* range brought about substantial activation of partially purified NADP-MDH from pea. Such concentrations do occur *in vivo* in the chloroplast.[5] Further work is needed to evaluate the importance of this mechanism for the regulation of NADP-MDH *in vivo*.

B. EFFECTS ON GENE EXPRESSION

Glutathione was recently shown to act as an elicitor of the transcription of mRNAs coding for the enzymes of the phenylalanine ammonia lyase (PAL) phytoalexin pathway.[150] Only the reduced form of glutathione was found to act as an elicitor, and other reducing agents had no effect. Similar results were reported for accumulation of the pterocarpin phytoalexin pisatin in pea.[151] Also, GSH enhances the dose-response of coumarin biosynthesis to elicitation by chitosan in parsley cells in conditioned growth medium.[152] This is a most intriguing result in light of the many responses to oxidative stress which are known to involve induction mechanisms. It may be that the increase in glutathione levels which follows the onset of the stress stimulus in so many instances has the consequence, not only of protecting the cell against the generation of free radicals by virtue of its own properties, but also of switching on the whole panoply of stress-resistance responses.

Edwards et al.[56] suggest that changes in GSH may be an effect rather than a cause of the elicitation response. In their hands, intracellular GSH levels could be manipulated (for example, by application of oxothiazolidine 4-carboxylate, which stimulates GSH biosynthesis) without bringing about PAL induction. They propose that since elicitation involves lipid peroxidation,[153] the increases in GSH observed may be the well-known inductive response to oxidative stress. However, the accumulated data do suggest that GSH alone, in the absence of elicitor, can bring about PAL induction. This apparent contradiction remains to be resolved.

GSSG has been known for some time to inhibit protein synthesis in bacterial and animal systems through an interaction with one of the initiation factors for translation.[154,155] It appears, now, that it has the same effect in a plant system, as has been recently demonstrated by Dhindsa et al.[156] An accumulation of GSSG may result, therefore, in deleterious effects on developmental responses or on those stress-resistance processes which involve induction mechanisms.

VII. CONCLUSIONS

Multiple regulatory "circuitry" may be involved in the sensing and processing of oxidative stress signals, circuitry which acts to trigger distinct but overlapping subsets of stress-resistance genes. The process is one which is critical for directing antioxidant capacity to specific subcellular compartments. The time course of stress responses suggests that events associated with biosynthesis of glutathione, in particular, may act as triggers or signals for the altered gene expression which follows. Glutathione biosynthesis is most likely a stromal event, and this is one of the first detectable metabolic events in stress responses. Yet the relationship of changes such as these occurring in the plastid to those taking place in other subcellular compartments has received little attention. The search for a stress response signal could perhaps start within the plastid with early responses of the glutathione biosynthetic pathway. A study of transport processes taking place at the chloroplast envelope could also yield important information since, whatever the signal is, it must leave the plastid to influence nuclear gene expression. By taking this direction, a step has been taken toward understanding the relationship between increased glutathione production and the induced nuclear gene expression which almost invariably seems to follow. Studies which focus on the behavior of antioxidants, such as glutathione within individual subcellular compartments, are essential if mechanisms underlying the adaptation of plant cells to oxidative stress are to be uncovered.

The stromal scavenging pathway functions as an antioxidant defense mechanism in plant chloroplasts, and confers resistance against oxidative stress regardless of its origin. Differences between biotypes with regard to resistance suggest the pleiotropic control of at least three of the enzymes involved. By analogy with bacterial systems, a regulatory gene may coordinate the expression of genes for stromal AP, SOD, and GR. Identification of the mechanism that triggers coordinate elevation of the scavenging cycle enzymes is clearly of value for manipulating the effectiveness of plant survival to oxidative stresses.

How plants detect oxidative stress is unknown. Results to date are compatible with the hypothesis that glutathione or other molecules synthesized under conditions of oxidative stress may stimulate the production of antioxidant enzymes.[157] While superoxide is highly reactive and most likely does not traverse the chloroplast membrane, glutathione and hydrogen peroxide are two plastid components whose accumulation in the chloroplast and potential to traverse subcellular membranes could affect changes in the nucleus.

REFERENCES

1. **Rennenberg, H.,** Glutathione metabolism and possible biological roles in higher plants, *Phytochemistry,* 21, 2771, 1982.
2. **Klapheck, S.,** Homoglutathione: isolation, quantification and occurrence in legumes, *Physiol. Plant.,* 74, 727, 1988.
3. **Kosower, E. M.,** Structure and reactions of thiols with special emphasis on glutathione, in *Coenzymes and Cofactors, Vol. III, Glutathione. Chemical, Biochemical, and Medical Aspects, Part A,* Dolphin, D., Poulson, R., and Avramovic, O., Eds., John Wiley & Sons, New York, 1989, 103.
4. **Mannervik, B., Carlberg, I., and Larson, K.,** Glutathione: general review of mechanism of action, in *Coenzymes and Cofactors, Vol. III, Glutathione. Chemical, Biochemical, and Medical Aspects, Part A,* Dolphin, D., Poulson, R., and Avramovic, O., Eds., John Wiley & Sons, New York, 1989, 475.
5. **Foyer, C. H. and Halliwell, B.,** The presence of glutathione and glutathione reductase in chloroplasts: a proposed role in ascorbic acid metabolism, *Planta,* 133, 21, 1976.
6. **Bielawski, W. and Joy, K. W.,** Reduced and oxidized glutathione and glutathione reductase activity in tissues of *Pisum sativum, Planta,* 169, 267, 1986.
7. **Law, M. Y., Charles, S. A., and Halliwell, B.,** Glutathione and ascorbic acid in spinach (*Spinacia oleracea*) chloroplasts. The effect of hydrogen peroxide and paraquat, *Biochem. J.,* 210, 899, 1983.
8. **Gillham, D. J. and Dodge, A. D.,** Hydrogen peroxide-scavenging systems within pea chloroplasts. A quantitative study, *Planta,* 167, 246, 1986.
9. **Klapheck, S., Latus, C., and Bergmann, L.,** Localization of glutathione synthetase and distribution of glutathione in leaf cells of *Pisum sativum* L., *J. Plant Physiol.,* 131 123, 1987.
10. **Smith, I. K., Kendall, A. C., Keys, A. J., Turner, J. C., and Lea, P. J.,** The regulation of the biosynthesis of glutathione in leaves of barley *(Hordeum vulgare* L.), *Plant Sci.,* 41, 11, 1985.
11. **Anderson, M. E.,** Enzymatic and chemical methods for the determination of glutathione, in *Coenzymes and Cofactors, Vol. III, Part A, Glutathione. Chemical, Biochemical, and Medical Aspects,* Dolphin, D., Poulson, R., and Avramovic, O., Eds., John Wiley & Sons, New York, 1989, 339.
12. **Fahey, R.,** Methods for determination of glutathione and derivatives using high-performance liquid chromatography, in *Coenzymes and Cofactors, Vol. III, Part A, Glutathione. Chemical, Biochemical, and Medical Aspects,* Dolphin, D., Poulson, R., and Avramovic, O., Eds., John Wiley & Sons, New York, 1989, 303.
13. **Loewus, F. A.,** Ascorbic acid and its metabolic products, in *The Biochemistry of Plants, Vol. 14,* Academic Press, New York, 1988, 85.
14. **Hossain, M. A., Nakano, Y., and Asada, K.,** Monodehydroascorbate reductase in spinach chloroplasts and its participation in regeneration of ascorbate for scavenging hydrogen peroxide, *Plant Cell Physiol.,* 25, 385, 1984.
15. **Foyer, C. H. and Halliwell, B.,** Purification and properties of dehydroascorbate reductase from spinach leaves, *Phytochemistry,* 16, 1347, 1977.
16. **Shigeoka, S., Yasumoto, R., Onishi, T., Nakano, Y., and Kitaoka, S.,** Properties of monodehydroascorbate reductase and dehydroascorbate reductase and their participation in the regeneration of ascorbate in *Euglena gracilis, J. Gen. Microbiol.,* 133, 227, 1987.

17. **Hausladen, A. and Kunert, K. J.,** Effects of artificially enhanced levels of ascorbate and glutathione on the enzymes monodehydroascorbate reductase, dehydroascorbate reductase, and glutathione reductase in spinach *(Spinacia oleracea), Physiol. Plant.*, 79, 384, 1990.
18. **Smith, I. K.,** Stimulation of glutathione synthesis in photorespiring plants by catalase inhibitors, *Plant Physiol.*, 79, 1044, 1985.
19. **Smith, I. K., Vierheller, T. L., and Thorne, C. A.,** Properties and functions of glutathione reductase in plants, *Physiol. Plant.*, 77, 449, 1989.
20. **Schirmer, R. H. and Krauth-Siegel, R. L.,** Glutathione reductase, in *Coenzymes and Cofactors, Vol. III, Part A, Glutathione. Chemical, Biochemical, and Medical Aspects,* Dolphin, D., Poulson, R., and Avramovic, O., Eds., John Wiley & Sons, New York, 1989, 553.
21. **Perry, A. C. F., Bhriain, N. N., Brown, N. L., and Rouch, D. A.,** Molecular characterization of the *gor* gene encoding glutathione reductase from *Pseudomonas aeruginosa:* determinants of substrate specificity among pyridinenucleotide- disulphide oxidoreductases, *Mol. Microbiol.*, 5, 163, 1991.
22. **Ermler, U. and Schulz, G. E.,** The three-dimensional structure of glutathione reductase from *Escherichia coli* at 3.0 Å resolution, *Proteins,* 9, 174, 1991.
23. **Hausladen, A. and Alscher, R. G.,** in preparation.
24. **Anderson, J. V., Hess, J. L., and Chevone, B. I.,** Purification, characterization, and immunological properties for two isoforms of glutathione reductase from Eastern white pine needles, *Plant Physiol.*, 94, 1402, 1990.
25. **Kalt-Torres, W., Burke, J. J., and Anderson, J. M.,** Chloroplast glutathione reductase: purification and properties, *Physiol. Plant.*, 61, 271, 1984.
26. **Mahan, J. R. and Burke, J. J.,** Purification and characterization of glutathione reductase from corn mesophyll chloroplasts, *Physiol. Plant.*, 71, 352, 1987.
27. **Connell, J. P. and Mullet, J. E.,** Pea chloroplast glutathione reductase: purification and characterization, *Plant Physiol.*, 82, 351, 1986.
28. **Halliwell, B. and Foyer, C. H.,** Properties and physiological function of a Glutathione reductase purified from spinach leaves by affinity chromatography, *Planta,* 139, 9, 1978.
29. **Edwards, E. A., Rawsthorne, S., and Mullineaux, P. M.,** Subcellular distribution of multiple forms of glutathione reductase in leaves of pea *(Pisum sativum* L.), *Planta,* 180, 278, 1990.
30. **Wise, R. R. and Naylor, A. W.,** Chilling-enhanced photooxidation. Evidence for the role of singlet oxygen and superoxide in the breakdown of pigments and endogenous antioxidants, *Plant Physiol.*, 83, 278, 1987.
31. **Shigeoka, S., Onishi, T., Nakano, Y., and Kitaoka, S.,** Characterization and physiological function of glutathione reductase in *Euglena gracilis z, Biochem. J.*, 242, 511, 1987.
32. **Guy, C. L. and Carter, J. V.,** Characterization of partially purified glutathione reductase from cold-hardened and nonhardened spinach leaf tissue, *Cryobiology,* 21, 454, 1984.
33. **Drumm-Herrel, H., Gerhäuber, U., and Mohr, H.,** Differential regulation by phytochrome of the appearance of plastidic and cytoplasmatic isoforms of glutathione reductase in mustard (*Sinapis alba* L.) cotyledons, *Planta,* 178, 103, 1989.
34. **Madamanchi, N. R., Anderson, J. V., Hess, J. L., and Alscher, R. G.,** in preparation.
35. **Foyer, C. H.,** in preparation.
36. **Liang, X., Dron, M., Cramer, C. L., Dixon, R. A., and Lamb, C. J.,** Differential regulation of phenylalanine ammonia-lyase genes during plant development and by environmental cues, *J. Biol. Chem.*, 264, 14486, 1989.
37. **Foyer, C. H., Dujardyn, M., and Lemoine, Y.,** Responses of photosynthesis and the xanthophyll and ascorbate-glutathione cycles to changes in irradiance, photoinhibition and recovery, *Plant Physiol. Biochem.*, 27, 751, 1989.
38. **Cséke, C. and Buchanan, B. B.,** Regulation of the formation and utilization of photosynthate in leaves, *Biochim. Biophys. Acta,* 853, 43, 1986.
39. **Bielawski, W. and Joy, K. W.,** Properties of glutathione reductase from chloroplasts and roots of pea, *Phytochemistry,* 25, 2261, 1986.
40. **Smith, I. K., Vierheller, T. L., and Thorne, C. A.,** Assay of glutathione reductase in crude tissue homogenates using 5,5′-dithiobis(2-nitrobenzoic acid), *Anal. Biochem.*, 175, 408, 1988.

41. **Wingsle, G.,** Purification and characterization of glutathione reductase from Scots pine needles, *Physiol. Plant.*, 76, 24, 1989.
42. **Mata, A. M., Pinto, M. C., and Lopez-Barea, J.,** Redox interconversion of glutathione reductase from *Escherichia coli*. A study with pure enzyme and cell-free extracts, *Mol. Cell Biochem.*, 67, 65, 1985.
43. **Meister, A.,** Glutathione metabolism and its selective modification, *J. Biol. Chem.*, 263, 17205, 1988.
44. **Hell, R. and Bergmann, L.,** Gamma-glutamylcysteine synthetase in higher plants: catalytic properties and subcellular localization, *Planta,* 180, 603, 1990.
45. **Hell, R. and Bergmann, L.,** Glutathione synthetase in tobacco suspension cultures: catalytic properties and localization, *Physiol. Plant.*, 72, 70, 1988.
46. **Law, M. Y. and Halliwell, B.,** Purification and properties of glutathione synthetase from spinach (*Spinacia oleracea*) leaves, *Plant Sci.*, 43, 185, 1986.
47. **Macnicol, P. K.,** Homoglutathione and glutathione synthetase of legume seedlings: partial purification and substrate specificity, *Plant Sci.*, 53, 229, 1987.
48. **Klapheck, S., Zopes, H., Levels, H. G., and Bergmann, L.,** Properties and localization of the homoglutathione synthetase from *Phaseolus coccineus* leaves, *Physiol. Plant.*, 74, 733, 1988.
49. **Rennenberg, H.,** Aspects of glutathione function and metabolism in plants, in *Plant Molecular Biology. NATO Advanced Studies Institute,* Wettstein, D. and Chua, N. H., Eds., Plenum Press, New York, 1987, 279.
50. **Steinkamp, R. and Rennenberg, H.,** Gamma-glutamyltranspeptidase in tobacco suspension cultures: catalytic properties and subcellular localization, *Physiol. Plant.*, 61, 251, 1984.
51. **Rennenberg, H., Steinkamp, R., and Kesselmeier, J.,** 5-Oxo-prolinase in *Nicotiana tabaccum:* catalytic properties and subcellular localization, *Physiol. Plant.*, 1981.
52. **Steinkamp, R., Schweihofen, B., and Rennenberg, H.,** Gamma-Glutamylcyclotransferase in tobacco suspension cultures: catalytic properties and subcellular localization, *Physiol. Plant.*, 69, 499, 1987.
53. **Rüegsegger, A., Schmutz, D., and Brunold, C.,** Regulation of glutathione biosynthesis by cadmium in *Pisum sativum* L, *Plant Physiol.*, 93, 1579, 1990.
54. **Buwalda, F., Stulen, I., De Kok, L. J., and Kuiper, P. J. C.,** Cysteine, gamma-glutamylcysteine and glutathione contents of spinach leaves as affected by darkness and application of excess sulfur. II. Glutathione accumulation in detached leaves exposed to H_2S in the absence of light is stimulated by the supply of glycine to the petiole, *Physiol. Plant.*, 80, 196, 1990.
55. **Hilton, J. L. and Pillai, P.,** L-2-oxothiazolidine-4-carboxylic acid protection against tridiphane toxicity, *Weed Sci.*, 34, 669, 1986.
56. **Edwards, R., Blount, J. W., and Dixon, R. A.,** Glutathione and elicitation of the phytoalexin response in legume cell cultures, *Planta,* 184, 403, 1991.
57. **Ahmad, T. and Frischer, H.,** Active site-specific inhibition by 1,3-*bis*(2-chloroethyl)-1-nitrosourea of two genetically homologous flavoenzymes: glutathione reductase and lipoamide dehydrogenase, *J. Lab. Clin. Med.*, 105, 464, 1985.
58. **Alscher, R. G.,** Biosynthesis and antioxidant function of glutathione in plants, *Physiol. Plant.*, 77, 457, 1989.
59. **Schupp, R. and Rennenberg, H.,** Diurnal changes in the glutathione content of spruce needles (*Picea abies* L.), *Plant Sci.*, 57, 113, 1988.
60. **Robinson, J. M.,** Does O_2 photoreduction occur in chloroplasts *in vivo? Physiol. Plant.*, 72, 666, 1988.
61. **Buwalda, F., De Kok, L. J., Stulen, I., and Kuiper, P. J. C.,** Cysteine, gamma-glutamylcysteine and glutathione contents of spinach leaves as affected by darkness and application of excess sulfur, *Physiol. Plant.*, 74, 663, 1988.
62. **Lamoureux, G. L. and Rusness, D. G.,** The role of glutathione and glutathione-S-transferases in pesticide metabolism, selectivity, and mode of action in plants and insects, in *Coenzymes and Cofactors, Vol. III, Glutathione. Chemical, Biochemical, and Medical Aspects, Part B,* Dolphin, D., Poulson, R., and Avramovic, O., Eds., John Wiley & Sons, New York, 1989, 153.

63. **Flohé, L.,** The selenium enzyme glutathione peroxidase, in *Coenzymes and Cofactors, Vol. III, Glutathione. Chemical, Biochemical, and Medical Aspects, Part A,* Dolphin, D., Poulson, R., and Avramovic, O., Eds., John Wiley & Sons, New York, 1989, 643.
64. **Smith, J. and Shrift, A.,** Phylogenetic distribution of glutathione peroxidase, *Comp. Biochem. Physiol.,* 63B, 39, 1979.
65. **Polle, A., Chakrabarti, K., Schürmann, W., and Rennenberg, H.,** Composition and properties of hydrogen peroxide decomposing systems in extracellular and total extracts from needles of Norway spruce (*Picea abies* L., Karst.), *Plant Physiol.,* 94, 312, 1990.
66. **Dhindsa, R. S.,** Drought stress, enzymes of glutathione metabolism, oxidation injury, and protein and synthesis in *Tortula ruralis, Plant Physiol.,* 95, 648, 1991.
67. **Yokota, A., Shigeoka, S., Onishi, T., and Kitaoka, S.,** Selenium as inducer of glutathione peroxidase in low-CO_2-grown *Chlamydomonas reinhardtii, Plant Physiol.,* 86, 649, 1988.
68. **Overbaugh, J. M. and Fall, R.,** Characterization of a selenium-independent glutathione peroxidase from *Euglena gracilis, Plant Physiol.,* 77, 437, 1985.
69. **Kuroda, H., Sagisaka, S., Asada, M., and Chiba, K.,** Peroxide-scavenging systems during cold acclimation of apple callus in culture, *Plant Cell Physiol.,* 32, 635, 1991.
70. **Halliwell, B. and Gutteridge, J. M. C.,** *Free Radicals in Biology and Medicine,* Clarendon, Oxford, 1985, 340.
71. **Salin, M. L.,** Toxic oxygen species and protective systems of the chloroplast, *Physiol. Plant.,* 72, 681, 1988.
72. **Bennet, J.,** Regulation of photosynthesis by chloroplast protein phosphorylation, *Phil. Trans. R. Soc. Lond. B,* 302, 113, 1983.
73. **Asada, K.,** Formation and scavenging of superoxide in chloroplasts, with relation to injury by sulfur dioxide, *Res. Rep. Natl. Inst. Environ. Stud.,* 11, 165, 1980.
74. **Asada, K., Takahashi, M. A., Tanaka, K., and Nakano, Y.,** Formation of active oxygen and its fate in chloroplasts, in *Biochemical and Medical Aspects of Active Oxygen,* Hayashi, O. and Asada, K., Eds., University Park Press, Baltimore, 1977, 45.
75. **Daub, M. E. and Hangarter, R. P.,** Light-induced production of singlet oxygen and superoxide by the fungal toxin cercosporin, *Plant Physiol.,* 73, 855, 1983.
76. **Heath, R. L.,** Biochemistry of ozone attack on the plasma membrane of plant cells, *Rec. Adv. Phytochem.,* 21, 29, 1987.
77. **Tanaka, K., Furusawa, I., and Kondo, N.,** SO_2 tolerance of tobacco plants regenerated from paraquat-tolerant callus, *Plant Cell Physiol.,* 29, 743, 1988.
78. **Tanaka, K., Suda, Y., Kondo, N., and Sugahara, K.,** O_3 tolerance and the ascorbate-dependent H_2O_2 decomposing system in chloroplasts, *Plant Cell Physiol.,* 26, 1425, 1985.
79. **Alscher, R. and Amthor, J. S.,** The physiology of free-radical scavenging: maintenance and repair processes, in *Air Pollution and Plant Metabolism,* Schulte-Hostede, S., Darrall, N. M., Blank, L. W., and Wellburn, A. R., Eds., Elsevier, Essex, 1988, 94.
80. **Öquist, G.,** Seasonally induced changes in acyl lipids and fatty acids of chloroplast thylakoids of *Pinus silvestris.* A correlation between the level of unsaturation of monogalactosyldiglyceride and the rate of electron transport, *Plant Physiol.,* 69, 869, 1982.
81. **Schmidt, A. and Kunert, K. J.,** Lipid peroxidation in higher plants. The role of glutathione reductase, *Plant Physiol.,* 82, 700, 1986.
82. **Kunert, K. J. and Dodge, A. D.,** Herbicide-induced radical damage and antioxidative systems, in *Target Sites of Herbicide Action,* Böger, P. and Sandmann, G., Eds., CRC Press, Boca Raton, FL, 1989, 45.
83. **Halliwell, B.,** *Chloroplast Metabolism,* Clarendon Press, Oxford, 1984, 257.
84. **Matringe, M. and Scalla, R.,** Studies on the mode of action of acifluorfen-methyl in non-chlorophyllous soybean cells, *Plant Physiol.,* 86, 619, 1988.
85. **Vaughan, K. C. and Duke, S. O.,** *In situ* localization of the sites of paraquat action, *Plant Cell Environ.,* 6, 13, 1983.
86. **Winterbourne, C. C.,** Production of hydroxyl radicals from paraquat radicals and H_2O_2, *FEBS Lett.,* 128, 339, 1981.
87. **Babbs, C. F., Pham, J. A., and Coolbaugh, R. C.,** Lethal hydroxyl radical production in paraquat treated plants, *Plant Physiol.,* 90, 1267, 1989.

88. **Grimes, H. D., Perkins, K. K., and Boss, W. F.,** Ozone degrades into hydroxyl radical under physiological conditions. A spin trapping study, *Plant Physiol.*, 72, 1016, 1983.
89. **Kanofsky, J. R. and Sima, P.,** Singlet oxygen production from the reactions of ozone with biological molecules, *J. Biol. Chem.*, 266, 9039, 1991.
90. **Mehlhorn, H., Tabner, B. J. and Wellburn, A. R.,** Electron spin resonance evidence for the formation of free radicals in plants exposed to ozone, *Physiol. Plant.*, 79, 377, 1990.
91. **Asada, K. and Kiso, K.,** Initiation of aerobic oxidation of sulfite by illuminated spinach chloroplasts, *Eur. J. Biochem.*, 33, 253, 1973.
92. **Sevilla, F., Lopez-Gorge, J., and Del Rio, L. A.,** Characterization of a manganese superoxide dismutase from the higher plant *Pisum sativum*, *Plant. Physiol.*, 70, 1321, 1982.
93. **Hayakawa, T., Kanematsu, S., and Asada, K.,** Occurrence of Cu,Zn-superoxide dismutase in the intrathylakoid space of spinach chloroplasts, *Plant Cell Physiol.*, 25, 883, 1984.
94. **Charles, S. A. and Halliwell, B.,** Effect of hydrogen peroxide on spinach (*Spinacia oleracea*) chloroplast fructose bisphosphatase, *Biochem. J.*, 189, 373, 1980.
95. **Groden, D. and Beck, E.,** H_2O_2 destruction by ascorbate-dependent systems from chloroplasts, *Biochim. Biophys. Acta*, 546, 426, 1979.
96. **Nakano, Y. and Asada, K.,** Spinach chloroplasts scavenge hydrogen peroxide on illumination, *Plant Cell Physiol.*, 21, 1295, 1980.
97. **Nakano, Y. and Asada, K.,** Hydrogen peroxide is scavenged by ascorbate-specific peroxidase in spinach chloroplasts, *Plant Cell Physiol.*, 22, 867, 1981.
98. **Hossain, M. A. and Asada, K.,** Inactivation of ascorbate peroxidase in spinach chloroplasts on dark addition of hydrogen peroxide: its protection by ascorbate, *Plant Cell Physiol.*, 25, 1285, 1984.
99. **Dalton, D. A., Hanus, F. J., Russell, S. A., and Evans, H. J.,** Purification, properties, and distribution of ascorbate peroxidase in legume root nodules, *Plant Physiol.*, 83, 789, 1987.
100. **Dalton, D. A., Russell, S. A., Hanus, F. J., Pascoe, G. A., and Evans, H. J.,** Enzymatic reactions of ascorbate and glutathione that prevent peroxide damage in soybean root nodules, *Proc. Natl. Acad. Sci. U.S.A.*, 83, 3811, 1986.
101. **Beauchamp, C. O. and Fridovich, I.,** Isozymes of superoxide dismutase from wheat germ, *Biochim. Biophys. Acta*, 317, 50, 1973.
102. **Young, L. C. T. and Conn, E. E.,** The reduction and oxidation of glutathione by plant mitochondria, *Plant Physiol.*, 31, 205, 1956.
103. **Leung, H. W., Vang, M. J., and Mavis, R. D.,** The cooperative interaction between Vitamin E and Vitamin C in suppression of peroxidation of membrane phospholipids, *Biochim. Biophys. Acta*, 664, 266, 1981.
104. **Barclay, L. R. C.,** The cooperative antioxidant role of glutathione with a lipid-soluble and a water-soluble antioxidant during peroxidation of liposomes initiated in the aqueous phase and in the lipid phase, *J. Biol. Chem.*, 263, 16138, 1988.
105. **Storz, G., Tartaglia, L. A., and Ames, B. N.,** Transcriptional regulator of oxidative stress-inducible genes: direct activation of oxidation, *Science*, 248, 189, 1990.
106. **Schmidt, A. and Kunert, K. J.,** Antioxidative defense systems: defense against oxidative damage in plants, in *Molecular Strategies for Plant Protection. UCLA Symposia on Molecular and Cellular Biology*, Arntzen, C. J. and Ryan, C., Eds., Alan R. Liss, New York, 1987, 401.
107. **Tanaka, K., Saji, H., and Kondo, N.,** Immunological properties of spinach glutathione reductase and inductive biosynthesis of the enzyme with ozone, *Plant Cell Physiol.*, 29, 637, 1988.
108. **Madamanchi, N. R., Alscher, R. G., and Cramer, C. L.,** in preparation.
109. **Tepperman, J. M. and Dunsmuir, P.,** Transformed plants with elevated levels of chloroplastic SOD are not more resistant to superoxide toxicity, *Plant Mol. Biol.*, 14, 501, 1990.
110. **Alscher, R., Franz, M., and Jeske, C. W.,** Sulfur dioxide and chloroplast metabolism, *Rec. Adv. Phytochem.*, 21, 1, 1987.
111. **Alscher, R., Bower, J., and Zipfel, W.,** The basis for different sensitivities of photosynthesis to SO_2 in two cultivars of pea, *J. Exp. Bot.*, 38, 99, 1987.
112. **Grill, D., Esterbauer, H., and Welt, R.,** Einflub von SO_2 auf das Ascorbinsäuresystem der Fichtennadeln, *Phytopath. Z.*, 96, 361, 1979.

113. **Grill, D., Esterbauer, H., and Hellig, K.,** Further studies on the effect of SO_2-pollution on the sulfhydril-system of plants, *Phytopath. Z.*, 104, 264, 1982.
114. **Chiment, J. J., Alscher, R., and Hughes, P. R.,** Glutathione as an indicator of SO_2-induced stress in soybean, *Env. Exp. Bot.*, 26, 147, 1986.
115. **Nieto-Sotelo, J. and Ho, T. D.,** Effect of heat shock on the metabolism of glutathione in maize roots, *Plant Physiol.*, 82, 1031, 1986.
116. **Mehlhorn, H., Seufert, G., Schmidt, A., and Kunert, K. J.,** Effect of SO_2 and O_3 on production of antioxidants in conifers, *Plant Physiol.*, 82, 336, 1986.
117. **Mehlhorn, H., Cottam, D. A., Lucas, P. W., and Wellburn, A. R.,** Induction of ascorbate peroxidase and glutathione reductase activities by interactions of mixtures of air pollutants, *Free Rad. Res. Comm.*, 3, 193, 1987.
118. **Jana, S. and Choudhuri, M. A.,** Effects of antioxidants on senescence and hill activity in three submerged plants, *Aquatic Bot.*, 27, 203, 1987.
119. **Chiment, J. and Hughes, P. R.,** personal communication.
120. **Heck, W. W. and Dunning, J. A.,** The effects of ozone on tobacco and pinto beans as conditioned by several ecological factors, *J. Air Poll. Control Assoc.*, 17, 112, 1967.
121. **Smith, I. K., Kendall, A. C., Keys, A. J., Turner, J. C., and Lea, P. J.,** Increased levels of glutathione in a catalase-deficient mutant of barley (*Hordeum vulgare* L.), *Plant Sci. Lett.*, 37, 29, 1984.
122. **Smirnoff, N. and Colombé, S. V.,** Drought influences the activity of enzymes of the chloroplast peroxide scavenging system, *J. Exp. Bot.*, 39, 1097, 1988.
123. **Schöner, S. and Krause, G. H.,** Protective systems against active oxygen species in spinach: response to cold acclimation in excess light, *Planta*, 180, 383, 1990.
124. **Shaaltiel, Y. and Gressel, J.,** Multienzyme oxygen radical detoxifying system correlated with paraquat resistance in *Conyza bonariensis*, *Pest. Biochem. Physiol.*, 25, 22, 1986.
125. **Shaaltiel, Y. and Gressel, J.,** Kinetic analysis of paraquat resistance in *Conyza:* evidence that paraquat transiently inhibits leaf chloroplast reactions in resistant plants, *Plant Physiol.*, 85, 869, 1987.
126. **Fuerst, E. P., Nakatani, H. Y., Dodge, A. D., Penner, D., and Arntzen, C. J.,** Paraquat resistance in *Conyza*, *Plant Physiol.*, 77, 984, 1985.
127. **Jansen, M. A. K., Malan, C., Shaaltiel, Y., and Gressel, J.,** Mode of photooxidant resistance to herbicides and xenobiotics, *Z. Naturforsch.*, 45c, 463, 1990.
128. **Shaaltiel, Y., Glazer, A., Bocion, P. F., and Gressel, J.,** Cross tolerance to herbicidal and environmental oxidants of plant biotypes tolerant to paraquat, sulfur dioxide and ozone, *Pest. Biochem. Physiol.*, 31, 13, 1988.
129. **Shaaltiel, Y. and Gressel, J.,** Biochemical analysis of paraquat resistance in *Conyza* leads to pinpointing synergists for oxidant generating herbicides, in *Pesticide Science and Biotechnology*, Greenhalgh, R. and Roberts, T. R., Eds., Blackwell Scientific, Oxford, 1987, 183.
130. **Gressel, J. and Shaaltiel, Y.,** Biorational herbicide synergists, in *Biotechnical Approaches to Plant Protection*, Hedin, P. A., Ed., Amer. Chem. Soc. Symp. Ser., Washington, D.C., 1991, 4.
131. **Burke, J. J., Gamble, P. E., Hatfield, J., and Quisenberry, J. E.,** Plant morphological and biochemical responses to field water deficits, *Plant Physiol.*, 79, 415, 1985.
132. **Clare, D. A., Rabinowitch, H. D., and Fridovich, I.,** Superoxide dismutase and chilling injury in *Chlorella ellipsoidea*, *Arch. Biochem. Biophys.*, 231, 158, 1984.
133. **Malan, C., Greyling, M. M., and Gressel, J.,** Correlation between CuZn superoxide dismutase and glutathione reductase, and environmental and xenobiotic stress tolerance in maize inbreds, *Plant Sci.*, 69, 157, 1990.
134. **Shaaltiel, Y., Chua, N.-H., Gepstein, S., and Gressel, J.,** Dominant pleiotropy controls enzymes co-segregating with paraquat resistance in *Conyza bonariensis*, *Theor. Appl. Genet.*, 75, 850, 1988.
135. **Hodgson, R. A. J. and Raison, J. K.,** Superoxide production by thylakoids during chilling and its implication in the susceptibility of plants to chilling-induced photoinhibition, *Planta*, 183, 222, 1991.

136. **Esterbauer, H. and Grill, D.,** Seasonal variation of glutathione and glutathione reductase in needles of *Picea abies, Plant Physiol.*, 61, 119, 1978.
137. **deKok, L. J. and Oosterhuis, F. A.,** Effects of frost hardening and salinity on glutathione and sulfhydryl levels and on glutathione reductase activity in spinach leaves, *Physiol. Plant*, 58, 47, 1983.
138. **Anderson, J. V., Chevone, B. I., and Hess, J. L.,** Seasonal changes in antioxidants of Eastern white pine, *Plant Physiol.*, in press.
139. **Hausladen, A., Doulis, A., Chevone, B. I., and Alscher, R. G.,** Glutathione reductase and cold tolerance of red spruce, in *Active Oxygen/Oxidative Stress and Plant Metabolism*, Pell, E. and Steffen, K., Eds., American Society of Plant Physiologists, 1991, 241.
140. **Mohapatra, S. S., Poole, R. J., and Dhindsa, R. S.,** Changes in protein patterns and translatable messenger RNA populations during cold acclimation of Alfalfa, *Plant Physiol.*, 84, 1172, 1987.
141. **Madamanchi, N. R. and Alscher, R. G.,** Metabolic bases for differences in sensitivity of two pea cultivars to sulfur dioxide, *Plant Physiol.*, 97, 88, 1991.
142. **Sen Gupta, A., Alscher, R. G., and McCune, D. C.,** Response of photosynthesis and cellular antioxidants to ozone in Populus leaves, *Plant Physiol.*, 96, 650, 1991.
143. **Matters, G. L. and Scandalios, J. G.,** Synthesis of isozymes of superoxide dismutase in maize leaves in response to O_3, SO_2 and elevated O_2, *J. Exp. Bot.*, 38, 842, 1987.
144. **Foyer, C. H., Rowell, J., and Walker, D. A.,** Measurement of the ascorbate content of spinach leaf protoplasts and chloroplasts during illumination, *Planta*, 157, 239, 1983.
145. **Anderson, J. W., Foyer, C. H., and Walker, D. A.,** Light-dependent reduction of hydrogen peroxide by intact spinach chloroplasts, *Biochim. Biophys. Acta*, 724, 69, 1983.
146. **Aono, M., Kubo, A., Saji, H., Natori, T., Tanaka, K., and Kondo, N.,** Resistance to active oxygen toxicity of transgenic *Nicotiana tabacum* that expresses the gene for glutathione reductase from *Escherichia coli, Plant Cell Physiol.*, 32, 691, 1991.
147. **Foyer, C. H., Lelandais, M., Galap, C., and Kunert, K. J.,** Effects of elevated cytosolic glutathione reductase activity on the cellular glutathione pool and photosynthesis in leaves under normal and stress conditions, *Plant Physiol.*, 97, 863, 1991.
148. **Wolosiuk, R. A. and Buchanan, B. B.,** Thioredoxin and glutathione regulate photosynthesis in chloroplasts, *Nature*, 266, 565, 1977.
149. **Vivekanandan, M. and Edwards, G. E.,** Activation of NADP-malate dehydrogenase in C3 plants by reduced glutathione, *Photosynth. Res.*, 14, 113, 1987.
150. **Wingate, V. P. M., Lawton, M. A., and Lamb, C. J.,** Glutathione causes a massive and selective induction of plant defense genes, *Plant Physiol.*, 87, 206, 1988.
151. **Yamada, T., Hashimoto, H., Shiraishi, T., and Oku, H.,** Suppression of pisatin, phenylalanine ammonia lyase mRNA and chalcone synthase mRNA accumulation by a putative pathogenicity factor from the fungus *Mycosphaeralla pinodes, Mol. Plant Microbe Interact.*, 2, 256, 1991.
152. **Conrad, U., Domard, R., and Kauss, H.,** Chitosan-elicited synthesis of callose and of coumarin derivatives in parsley cell suspension cultures, *Plant Cell Rep.*, 8, 152, 1989.
153. **Rogers, K. R., Albert, F., and Anderson, A. J.,** Lipid peroxidation is a consequence of elicitor activity, *Plant Physiol.*, 86, 547, 1988.
154. **Fahey, R. C., Brody, S., and Mikolajczyk, S. D.,** Changes in the glutathione thiol-disulfide status of *Neurospora crassa* conidia during germination and aging, *J. Bact.*, 121, 144, 1975.
155. **Kan, B., London, I. M., and Levin, D. H.,** Role of reversing factor in the inhibition of protein synthesis initiation by oxidized glutathione, *J. Biol. Chem.*, 263, 15652, 1988.
156. **Dhindsa, R. S.,** Glutathione status and protein synthesis during drought and subsequent rehydration in Tortular ruralis, *Plant Physiol.*, 83, 816, 1987.
157. **Alscher, R. G., Madamanchi, N. R., and Cramer, C. L.,** Protective mechanisms in the chloroplast stroma, in *Active Oxygen/Oxidative Stress and Plant Metabolism*, Pell, E. and Steffen, K., Eds., American Society of Plant Physiologists, 1991, 145.
158. **Jansen, M. A. K., Shaaltiel, Y., Kazzes, D., Canaani, O., Malkin, S., and Gressel, J.,** Increased tolerance to photoinhibitory light in paraquat-resistant *Conyza bonariensis* measured by photoacoustic spectroscopy and $^{14}CO_2$-fixation, *Plant Physiol.*, 91, 1174, 1989.

159. **Harvey, B. M. R. and Harper, D. B.,** Tolerance to bipyridylium herbicides, in *Herbicide Resistance in Plants,* LeBaron, H. M. and Gressel, J., Eds., John Wiley & Sons, New York, 1982, 215.
160. **Huttunen, S. and Heiska, E.,** Superoxide dismutase (SOD) activity in Scots pine (*Pinus sylvestris* L.) and Norway spruce (*Picea abies* L. Karst.) needles in northern Finland, *Eur. J. For. Pathol.*, 18, 343, 1988.

Chapter 2

ASCORBIC ACID

Christine H. Foyer

TABLE OF CONTENTS

0-8493-6328-4/93/$0.00 + $.50

I. INTRODUCTION

High concentrations of L-ascorbic acid (vitamin C) are characteristic of plant tissues. Indeed, ascorbate is one of the most important vitamins in the human diet,[1] being obtained largely from vegetables, fruit, and other plant material. In spite of this importance, the pathway of ascorbate synthesis in plants and its subsequent roles in plant metabolism have not been fully elucidated.[2,3] It is clear that ascorbate has effects on many physiological processes. Ascorbate has been implicated in the regulation of growth, differentiation, and metabolism in plants, but the precise nature of these effects has not been clearly defined.[2] In the following discussion, it is suggested that in most, if not all, of ascorbate's affects in plant tissues it is in acting essentially as a reductant. Ascorbate reacts with and scavenges many types of free radicals, both in animal and plant tissues. It also has effects on many enzyme activities, but the actual role of ascorbate is often to keep metal ions associated with such enzymes in the reduced form.[1] Thus, ascorbate need not have an obligatory role in catalysis. The basis for the significance of ascorbate in plant biology may thus reside largely in its reducing ability, minimizing the damage caused by oxidative processes. However, it may be correct to assume that other functions of ascorbate remain to be discovered. Whether the effects of ascorbate on growth and in the ripening processes of fruit are due to the general antioxidant role of ascorbate or to more specific functions is, as yet, unclear.

Green leaves may contain as much ascorbate as chlorophyll.[4,5] The presence of such a remarkable amount of ascorbate may suggest considerable metabolic significance, particularly if the amount is related to biological importance. A major function of ascorbate in plant cells was identified at the end of the 1970s when the participation of ascorbate in the scavenging of hydrogen peroxide (H_2O_2) was demonstrated.[6-8] It subsequently became clear that plants have evolved a unique peroxidase using ascorbate as electron donor.[7-9] In this view, the presence of high concentrations of ascorbate in plant tissues may be explained in terms of the fundamental role of ascorbate in the plant defense systems that protect metabolic processes against H_2O_2 and other toxic derivatives of oxygen. As Henry Gee of the Nature-Times news service (1989) put it: "Ascorbate, by reacting with oxidizing agents much more readily than anything else, mops them up before they have a chance to damage anything".

It has only recently been appreciated that ascorbate is an important antioxidant in both animal and plant tissue.[1,6] In plants, its role in H_2O_2 detoxification is pivotal for the prevention of oxidative damage to cellular integrity and metabolism, and in particular, for photosynthesis. Photosynthetic carbon assimilation is dependent on the stromal H_2O_2-scavenging system for its continued functioning in the light. Photosynthesis is extremely sensitive to low concentrations of H_2O_2.[10] A 50% inhibition of photosynthesis has been observed with 10 μM H_2O_2.[10] Even under optimum conditions for photosynthesis in air, the production of H_2O_2 by the thylakoid membranes would be sufficient to inhibit photosynthetic carbon assimilation in less than a second.[11] Chloroplasts do not contain catalase. Catalase is localized in the peroxisomes of the plant cell where its role is to eliminate excess H_2O_2 produced therein during photorespiration and other processes. In addition, catalase is not particularly well suited to the natural state in leaves, since it can undergo a light-induced inhibition of function, particularly at low temperatures.[12] Catalase has a very high catalytic capacity but a very poor affinity for its substrate, H_2O_2 (measured Km values approach 1 *M*), and catalase cannot efficiently remove the low H_2O_2 levels that inhibit the thiol-modulated enzymes of the Benson-Calvin cycle.[10]

Glutathione peroxidase, which is important in H_2O_2 detoxification in animals is largely absent from plant tissues. A glutathione peroxidase from *Euglena gracilis* has been purified to homogeneity, and was found not to contain selenium (unlike the animal glutathione peroxidase).[13] This enzyme was a tetramer with a high affinity for its substrates.[13] Recently, a cDNA clone for glutathione peroxidase from plants has been isolated.[14] The level of expression of this gene was high in protoplasts isolated from *Nicotiana sylvestris* leaves; however, it was not found to be normally expressed in leaves, stems, or roots. Its induction in these tissues occurred only in response to extreme stress. The gene was expressed at a high level in germinating seeds, in flowers, and also in the apex under nonstressed conditions.[14] Homologies with several animal selenium-dependent glutathione peroxidases were found; however, this protein may not be selenium dependent.[14] Glutathione peroxidase activity is thus absent from leaf extracts; however, Jablonski and Anderson[15] showed that H_2O_2-dependent oxidation of glutathione could be measured in pea-shoot extracts, but this involved more than one protein. Clearly, most plant tissues use ascorbate-specific peroxidases, rather than glutathione peroxidase, for the efficient removal of low concentrations of H_2O_2 under nonstressed conditions.

In the light, the chloroplasts of higher plants produce several destructive oxygen-derived species, including superoxide ($O_2^{\cdot -}$), H_2O_2, and the hydroxyl radical ($OH^{\cdot}$). Ascorbate is the terminal electron donor in the processes which scavenge these free radicals in the hydrophilic environments of plant cells. Ascorbate scavenges hydroxyl radicals at diffusion-controlled rates.[16] It reacts with $O_2^{\cdot -}$ with a rate constant of $2.7 \times 10^{-5}\ M^{-1}\ s^{-1}$, as shown in Reaction 1:

$$2O_2^{\cdot -} + 2H^+ + \text{ascorbate} \longrightarrow \text{dehydroascorbate} + 2H_2O_2 \quad (1)$$

Nishikimi[17] suggested that the reaction rates of superoxide with ascorbate and with superoxide dismutase were comparable in tissues, or in organelles such as the chloroplasts, where the ascorbate content is in the millimolar range. Ascorbate also reacts nonenzymically with H_2O_2 at a significant rate, as shown in Reaction 2:

$$H_2O_2 + 2\ \text{ascorbate} \longrightarrow + 2H_2O + 2\ \text{monodehydroascorbate} \quad (2)$$

This reaction is catalyzed by the ascorbate-specific peroxidases (EC 1.11.1.7) in the chloroplasts and cytosol of higher plants.[9,18,19]

In addition to its role as a primary antioxidant, ascorbate has a significant secondary antioxidant function. The ascorbate pool represents a reservoir of antioxidant potential that is used to regenerate other membrane-bound antioxidants such as α-tocopherol and zeaxanthin.[20,21] These scavenge lipid peroxide radicals and singlet oxygen, respectively.

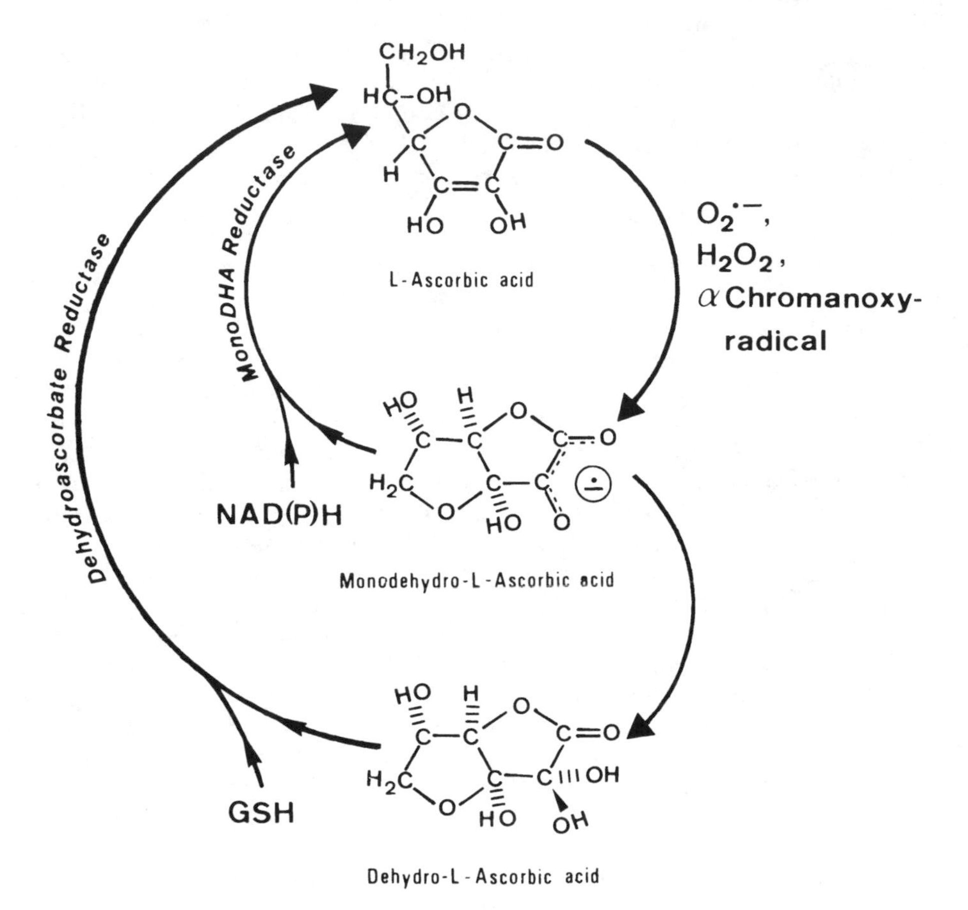

FIGURE 1. Interconversions between L-ascorbic acid and its oxidized forms monohydroascorbic acid and dehydroascorbate. Monodehydroascorbate (mono DHA), reduced glutathione (GSH).

Seasonal variations in the ascorbate content of leaves (particularly the needles of conifers) has been found; however, on the whole, the ascorbate pool in chloroplasts and leaves is maintained at a remarkably constant level.[22-24] Indeed, because ascorbate synthesis is often increased under stress conditions, the loss of the ascorbate pool may be a good indicator of the degree of stress experienced by plant tissues.[25] The size and reduction state of the ascorbate pool reflect the status of the leaf in terms of its antioxidant capacity, and thus can provide an approximation of the effects of stress on the photosynthetic apparatus. The functioning of ascorbate as an antioxidant is ensured by the action of the ascorbate-glutathione cycle that provides an efficient mechanism for recycling oxidized ascorbate and maintaining the ascorbate pool in its reduced form.[6,11]

Oxidation of ascorbic acid (midpoint potential for one electron donation = +0.330 V)[26] occurs in two sequential steps, forming, in the first instance, monodehydroascorbate, and subsequently dehydroascorbate (Figure 1). The monodehydroascorbate radical is the primary product of the ascorbate peroxidase reaction (Reaction 2), and is also produced by the univalent oxidation of ascorbate by superoxide (Reaction 1) and hydroxyl radicals, and by the tocopheroxyl (or α-chromanoxy) radical (Reaction 3) which is formed by the reaction of lipid peroxy-radicals with α-tocopherol:

$$\text{tocopheroxyl radical} + \text{ascorbate} \longrightarrow \alpha\text{-tocopherol} + \text{monodehydroascorbate} \quad (3)$$

Energetic and kinetic considerations of redox reactions of both ascorbate and α-tocopherol suggest that cycling between reduced and free radical forms occurs via the transfer of single hydrogen atoms rather than via separate electron transfer and protonation reactions.[26] At physiological pH, a process of hydrogen atom transfer in the free-radical scavenging activities of these vitamins would minimize deleterious events, such as direct reduction of molecular oxygen, and yet allow the vitamins to react efficiently with free radicals.[26] Hence, at physiological pH, ascorbate is a poor electron donor but a good donor of single hydrogen atoms.

The monodehydroascorbate radical spontaneously disproportionates to ascorbate and dehydroascorbate ($10^5\ M^{-1}\ s^{-1}$ at pH 7.0). Monodehydroascorbate radicals are also directly reduced to ascorbate by the action of NAD(P)H-dependent monodehydroascorbate reductase[11,27] (EC 1.6.5.4), as shown in Figure 1. Dehydroascorbate is also highly unstable at pH values greater than pH 6.0. The carbon chain is cleaved to products such as tartrate and oxalate,[3] and may decompose to yield toxic derivatives. Part of the oxalate formed in plant tissues appears to arise from ascorbate. To prevent loss of the ascorbate pool following oxidation, the chloroplast contains efficient mechanisms of recycling both monodehydroascorbate and dehydroascorbate, and these ensure that the ascorbate pool is maintained largely in the reduced form (Figure 1). Reduced glutathione (GSH), which is present in chloroplasts in millimolar concentrations, will nonenzymically reduce dehydroascorbate back to ascorbate at pH values greater than pH 7.0 (Reaction 4):

$$2\text{GSH} + \text{dehydroascorbate} \longrightarrow \text{GSSG} + \text{ascorbate} \quad (4)$$

However, this reaction is catalyzed by a dehydroascorbate reductase (GSH: dehydroascorbate oxidoreductase, E.C. 1.8.5.1) which is present at high activities in leaves, seeds, and other tissues.[28] The ubiquitous occurrence of dehydroascorbate reductase suggests that the rate of nonenzymatic reduction of dehydroascorbate is not adequate. Indeed, it may account for less than 0.1% of the enzyme catalyzed rate *in situ*.[29]

II. OCCURRENCE AND COMPARTMENTATION

In higher plants, the ascorbic acid content varies greatly between tissues. Organs and tissues undergoing active growth contain between 0.1 to 2 mg of ascorbic acid per gram fresh weight. Ascorbic acid is virtually undetectable in dry seeds, while leaves, inflorescences, and pollen contain substantial amounts. In certain fruits, it

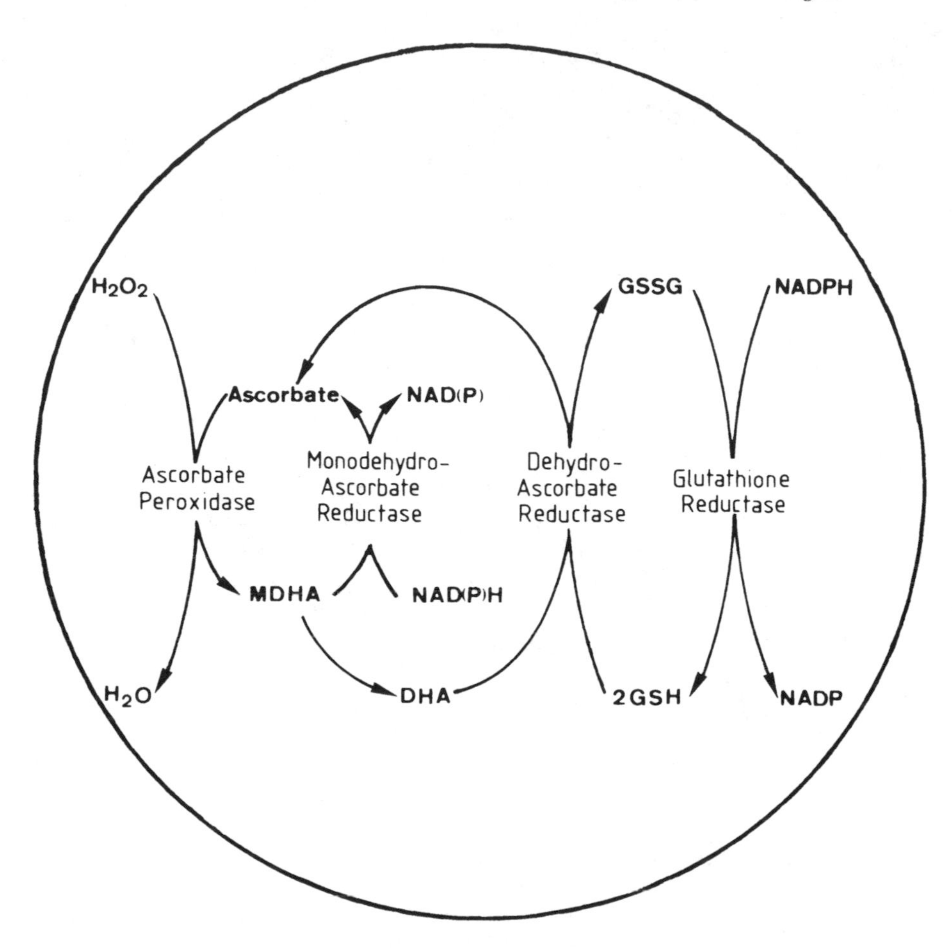

FIGURE 2. The ascorbate-glutathione cycle. Dehydroascorbate (DHA), reduced glutathione (GSH), oxidized glutathione (GSSG), monodehydroascorbate (MDHA).

accounts for about 0.1% of the fresh weight. The localization of ascorbate in nongreen tissues remains to be elucidated, but the intracellular distribution in green leaves has been studied. Leaf protoplasts contain 20 to 30% less ascorbate than the leaves from which they were obtained, suggesting that a significant proportion of the ascorbate in the leaf is present in the cell wall or apoplast.[30] Ascorbate, glutathione, and cysteine have been found in the apoplastic space of spruce needles at concentrations of 1.1 m*M*, 6.5 μ*M*, and 0.5 μ*M*, respectively.[31] These results suggest that ascorbate is a natural constituent of the apoplastic space, although its function in this location is not clear. Ascorbate peroxidase and the other enzymes of the ascorbate-glutathione cycle (Figure 2) are not found in the apoplastic space. However, ascorbate oxidase has been found to be associated with the cell wall in all tissues of *Cucurbita* species and also in cabbage leaves and barley roots,[32,33] and ascorbate may have a role in cell wall biosynthesis. Extra-cellular ascorbate

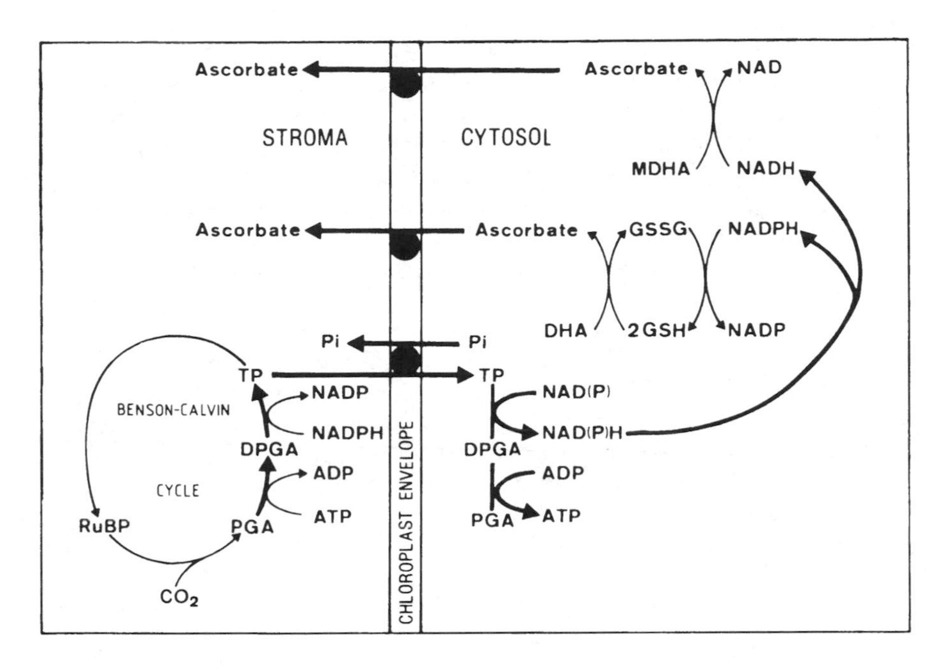

FIGURE 3. Scheme for the operation of the ascorbate-glutathione cycle in the chloroplast and cytosolic compartments of the leaf mesophyll cells showing the possible redox-links involved in the process between these compartments. 1,3-diphosphoglycerate (DPGA), dehydroascorbate (DHA), oxidized glutathione (GSSG), reduced glutathione (GSH), 3-phosphoglycerate (PGA), inorganic phosphate (Pi), ribulose-1,5-bisphosphate (RuBP), triose-phosphate (TP). (In this scheme, some of the leaf monodehydroascorbate reductase activity is considered to reside in the extra-chloroplastic fraction).

must be derived from within the cell and be transported across the plasmalemma membrane. In addition, since the apoplast does not have a system for regenerating reduced ascorbate, oxidized ascorbate or its oxidation products must be transported back into the cytosol. It is, therefore, possible that the plasmalemma of leaf cells contains an ascorbate translocator.

The ascorbate content of chloroplasts is generally between 10 and 50 m*M*, but values as high as 75 m*M* have been reported.[4,5,22] The chloroplasts contain only 20 to 40% of the ascorbate present in the mesophyll cells[22,30,34,35] and 10 to 50% of the leaf glutathione.[34,36] These results demonstrate that a high proportion of the leaf ascorbate and glutathione exists in the extra-chloroplast compartment of the leaf cell. Ascorbate is found both in the cytosol and vacuole; indeed, ascorbate may be synthesized outside of the chloroplast. However, this ascorbate is accessible to the chloroplast because the chloroplast envelope membrane contains a specific ascorbate translocator.[37,38] The ascorbate-glutathione cycle can operate efficiently in both the chloroplast and cytosol compartments with the transport of ascorbate and reducing equivalents (via the phosphate translocator) forming a redox link between the two compartments (Figure 3). While the ascorbate-glutathione cycle has been most extensively studied in chloroplasts, there is good evidence that all of the necessary enzymes of the cycle are also located in the cytosol of both green and nongreen tissues.[34,39,40] It is clear that there are both cytosolic and chloroplastic isoforms of ascorbate peroxidase.[41] The pea leaf cytosolic isoenzyme has been purified to homogeneity and characterized.[42] Anderson et al.[37] found only 30% of the total leaf dehydroascorbate reductase in spinach leaf chloroplasts, while Gillham and Dodge[34] found 65% of the total enzyme activity in pea chloroplasts. Both authors found 70% of the total glutathione reductase (EC 1.6.4.2) activity in the chloroplasts, as did Edwards et al.[43] These authors[43] found that 3% of the pea leaf extrachloroplastic glutathione reductase was localized in the mitochondria and 27% in the cytosol.

The occurrence of monodehydroascorbate reductase outside the chloroplast in plant tissues has not been reported. The enzyme has been purified to homogeneity from cucumber fruit.[44] The presence of a high NADH-dependent activity (twice that with NADPH) might suggest that an extrachloroplastic isoform may exist. In mammalian cells, this enzyme is associated with organelles such as the mitochondria, microsomes, and golgi.[45] With the exception of monodehydroascorbate reductase, all of the other enzymes of the ascorbate-glutathione cycle have been found to occur in the cytosol of green cells, as well as in the chloroplast. The ascorbate-dependent H_2O_2-scavenging system clearly functions in the cytosol of nongreen tissues,[39,40,46] and for the above reasons we may suggest that it is also functional in the cytosol of the leaf, as shown in Figure 3.

Since it has been found that some glutathione peroxidase is expressed in germinating seeds, in the apex, and in flowers, as is glutathione reductase, it is possible to speculate that a simple glutathione-dependent H_2O_2-scavenging system occurs in these tissues. However, the absence of glutathione peroxidase from leaves, stems, and roots[14] suggests that the ascorbate-glutathione cycle is the major defense system in these tissues.[15,47] The ascorbate oxidase activity of leaves is also low.

Seasonal variations occur in the ascorbate and glutathione contents of plant tissues. In addition, the activities of the enzymes of ascorbate-glutathione cycle vary with leaf age and the environmental conditions to which the plants are exposed.[48] The ascorbate content of leaves of maize plants produced in a growth chamber is shown in Figure 4. Clearly, the ascorbate content of leaves is related to their position on the plant, and may differ because of both leaf age and the irradiance arriving at the leaf surface. A diurnal rhythm in the glutathione concentration in spruce needles has been observed.[49] This rhythm was found to be light-dependent, but temperature independent, the highest concentrations occurring at midday, and the lowest at night. This diurnal rhythm was superimposed upon seasonal changes in the glutathione level. A similar diurnal rhythm has been observed in the ascorbate pool in spruce needles, but the changes in the pool size were much less (20% of the total pool).[50] It remains to be seen whether diurnal fluctuations in the ascorbate content occur in other plant species.

In the short term, irradiance has little effect on the ascorbate content of leaves or chloroplasts.[22-24,30] Figure 5 shows the effect of irradiance and CO_2 fixation on the ascorbate levels in isolated leaf protoplasts and chloroplasts. It is clear that the ascorbate pool is maintained at a remarkably constant level and redox state despite the change from darkness to light and regardless of whether CO_2 assimilation is functional or not.[22,30]

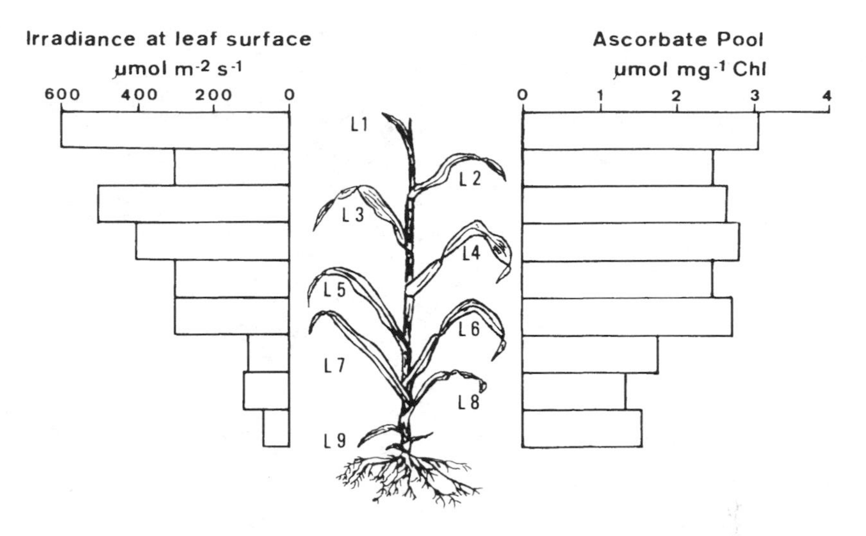

FIGURE 4. The effect of irradiance arriving at the leaf surface, and leaf position on the ascorbate contents of maize leaves in a growth chamber.

In the chloroplasts, most ascorbate is stromal, but a portion[30] (10 to 20%) is bound to the thylakoid membranes where it is involved in specific reactions such as the turnover of the xanthophyll cycle[21,51] and the regeneration of α-tocopherol. However, a pertinent question concerns the mechanism by which stromal ascorbate participates in thylakoid membrane and thylakoid lumen reactions. Ascorbate will not readily cross membranes without the aid of a carrier system. In contrast to the chloroplast envelope which affords ascorbate uptake into the stroma by a mechanism of facilitated diffusion,[37,38] all studies suggest that ascorbate transport into the thylakoid lumen is slow and occurs by diffusion alone with a half-time of 15 to 20 min. The apparent half-time for entry at 4°C was about 17 min at pH 7.5 in spinach thylakoids (Don Ort, personal communication).

III. METABOLISM

A. ASCORBIC ACID BIOSYNTHESIS

L-Ascorbic acid is a product of hexose metabolism in plants. The involvement of phosphorylated intermediates in the pathway of ascorbate synthesis has been demonstrated in algae,[52] but not in higher plants.[3,53-55] The pathway of ascorbic acid biosynthesis is still not fully understood and is problematic, since some points of controversy remain. Our present understanding of the process owes much to the dedicated research efforts of Frank Loewus.[3,53-57] Ascorbic acid synthesis cannot be directly compared to that found in animals because a branch point occurs in the reaction pathway when D-glucose is oxidized to the level of sugar acid in plant tissues. Two discrete biosynthetic pathways for the conversion of D-glucose to ascorbic acid are possible (Figure 6). Higher plants primarily use a direct pathway which conserves the carbon chain in the same order and carbon sequence as that

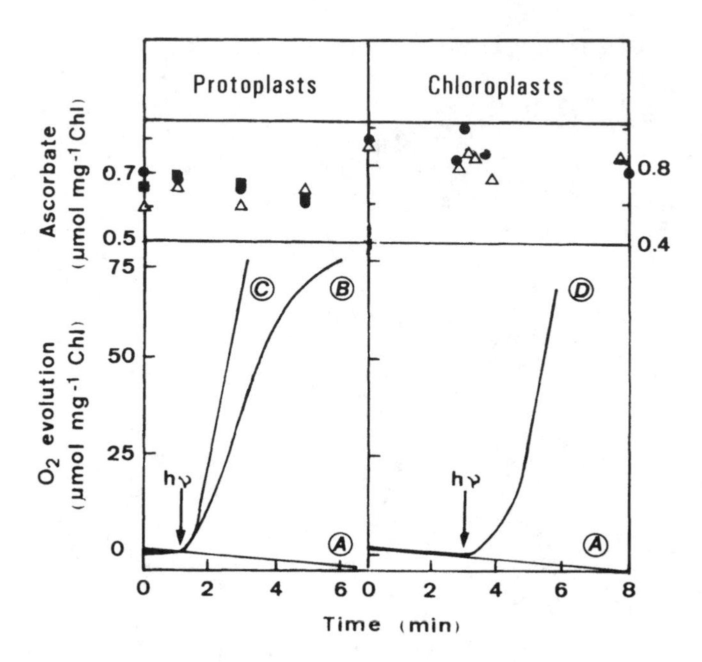

FIGURE 5. The effect of irradiance on the CO_2 assimilation rate (solid lines) and the ascorbate contents (open symbols) of spinach leaf protoplasts and chloroplasts. CO_2-dependent, O_2 evolution, and O_2 uptake in the absence of CO_2 were measured in intact chloroplasts and protoplasts in the oxygen electrode. Oxygen evolution or uptake in the absence of CO_2 (A), and the presence of either 0.25 m*M* (B) or 50 m*M* (C, D) $NaHCO_3$ is shown. The chloroplasts in this experiment had a high ascorbate content. They were prepared by mechanical techniques from a separate batch of leaves from those used for the protoplasts. This figure also illustrates that the level of ascorbate present can be quite variable depending on the original ascorbate level in the leaves, from which the cells and organelles were prepared.

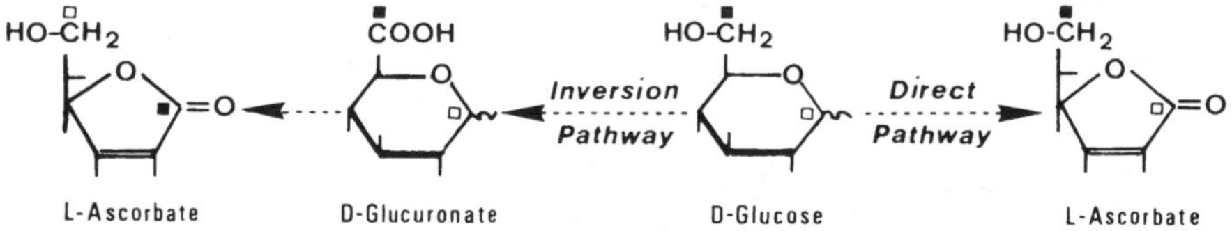

HO-CH2 COOH HO-CH2 HO-CH2

Inversion Pathway Direct Pathway

L-Ascorbate D-Glucuronate D-Glucose L-Ascorbate

FIGURE 6. The direct and inversion pathways of L-ascorbic acid synthesis. Symbols (□, ■) denote the terminal carbons of D-glucose. The inversion pathway proceeds through D-glucuronate and L-glucurnate in animals.

found in D-glucose. However, lower plants, for example one diatom[56] and two algal species,[52,57] have been shown to utilize the inversion pathway of ascorbate biosynthesis found in animals in which D-glucose is converted to the uronic acids and their derivatives by oxidation of carbon 6 (C-6), a reduction of C-1, and an enediol-generating oxidation. The numerical sequence of carbon atoms between glucose and ascorbic acid is, thus, inverted during this conversion (Figure 6). Humans and primates do not have the enzyme, L-gulono-γ-lactone oxidase, the last enzyme of the inversion pathway of ascorbic acid synthesis, and thus cannot form ascorbic acid from L-gulono-γ-lactone.

All of the available evidence tends to favor the operation of the direct pathway of ascorbic acid synthesis in higher plants, i.e., it does not involve "inversion". In this route, the C-1 of D-glucose is oxidized during ascorbate biosynthesis. The pathway of ascorbic acid biosynthesis is then considered to involve the oxidation of C-1 of D-glucose, an internal oxidation of C-2 or C-3, i.e., an enediol-generating oxidation and a retention of the hydroxy-methyl function at C-6.[3] A tentative scheme proposed by Loewus[3] is as follows:

$$\text{D-glucose} \longrightarrow \text{D-glucosone} \longrightarrow \text{L-sorbosone} \longrightarrow \text{L-ascorbic acid}$$

The conversion of glucosone and sorbosone to ascorbic acid has been observed in bean and spinach leaves.[54] These observations[54] suggest that glucosone is an intermediate formed from glucose, and an epimerization at carbon 5 converts glucose to sorbosone. An NADP-dependent dehydrogenase catalyzing the conversion of L-sorbosone to L-ascorbic acid has been isolated.[55] Interestingly, the alkaloid lycorine, which has been considered to be a potent inhibitor of ascorbic acid biosynthesis, has no effect on the dehydrogenase reaction.[55] The situation may, however, be more complicated, since higher plants readily convert L-galactono-γ-lactone to ascorbic acid. This substrate does not seem to be a natural constituent of plants, i.e., they do not apparently convert D-glucose to L-galactono-1,4-lactone. The presence of L-galactono-γ-lactone oxidase remains incompatible with the "direct" reaction scheme. It is possible that there is a natural substrate for this enzyme, as yet unknown, that arises from a noninversion pathway. This inconsistency with the direct pathway of ascorbic acid synthesis remains to be resolved. In addition, the intracellular localization of ascorbate synthesis has not been established.

B. THE ASCORBATE-GLUTATHIONE CYCLE AND H_2O_2 METABOLISM

Foyer and Halliwell[6] proposed that the redox pairs dehydroascorbate-ascorbate and oxidized glutathione-reduced glutathione (GSSG-GSH) are intermediate electron carriers in the reduction of H_2O_2 by NADPH in chloroplasts (Figure 2). The proposed role of the dehydroascorbate-ascorbate and GSSG-GSH redox pairs in the reduction of H_2O_2 was supported by changes in the concentrations of ascorbate and GSH, both in intact chloroplasts and ruptured chloroplasts in response to light and exogenously supplied H_2O_2.[37,47,58] Illuminated chloroplasts reduce O_2 to H_2O_2 using reducing equivalents from photosystem I in the process known as pseudocyclic electron flow (Figure 7). H_2O_2 is then metabolized using NADPH as the electron donor via the reaction sequence of the ascorbate-glutathione cycle, as illustrated in Figures 2 and 7. Isolated intact chloroplasts metabolize low concentrations of exogenously supplied H_2O_2 at rapid rates in the light, but not in the dark, with

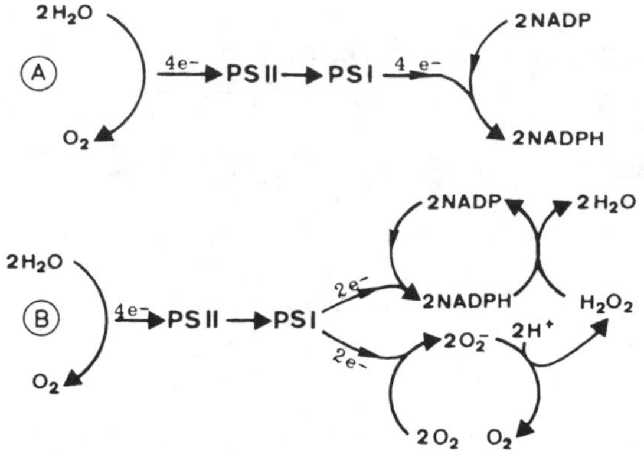

FIGURE 7. Diagrammatic representation of the paths of noncyclic electron flow (A) and noncyclic and pseudocyclic electron flow coupled to H_2O_2-scavenging (B) in the photosynthetic electron transport system. Photosystem I (PS I), Photosystem II (PS II).

concomittant evolution of O_2.[8,30,58] Thus, light-dependent H_2O_2 reduction to H_2O, as illustrated in Figure 7, may be written as in Reaction 5:

$$H_2O_2 + H_2O \longrightarrow 2H_2O + 1/2\ O_2 \quad (5)$$

H_2O_2 reduction is favored at alkaline pH values, e.g., optimum pH 8.2 in pea chloroplasts[47] and pH 8.5 in spinach chloroplasts.[59] Both O_2 evolution and H_2O_2 consumption are abolished in the presence of the electron transport inhibitor, 3-(3,4-dichlorophenyl)-1,1-dimethylurea, (DCMU).[58] The H_2O_2-scavenging system of the chloroplast is very efficient in the light, and has a high affinity for H_2O_2 in agreement with the Km values for H_2O_2 reported for chloroplast ascorbate peroxidase (50 to 80 μM).[9,41] The ability of the chloroplasts to metabolize H_2O_2 is lost if the chloroplasts are treated with the inhibitor sodium azide.[58] Furthermore, the addition of H_2O_2 to intact chloroplasts in the dark causes the inactivation of ascorbate peroxidase[58] because the ascorbate pool is rapidly oxidized by H_2O_2 and cannot be regenerated at a sufficiently rapid rate in the absence of electron transport which synthesizes NADPH. The chloroplast isoenzyme requires ascorbate for stability and becomes inactive in its absence.[9,11]

Ascorbate peroxidase activity in leaves appears to exist in compartment specific isoforms, the chloroplast isoform possibly being the most abundant.[18,19] Unlike the majority of plant peroxidases, ascorbate peroxidase has a high specificity for its substrate, ascorbate, the oxidation of guaicol being only 3% that of ascorbate. The enzyme catalyzes the reaction shown in Reaction 2 with a high affinity for both substrates, comparable to physiological concentration of hydrogen peroxide[11] and far below that of ascorbate.[22,23] The combined action of monodehydroascorbate reductase and dehydroascorbate reductase sustains the ascorbate pool predominantely in its reduced form *in situ*. Table 1 shows the relative activities of the component enzymes in leaves of wheat, pea, maize, and barley. While a high

TABLE 1
The Relative Activities of Enzymes of the Ascorbate-Glutathione Cycle in Pea, Maize, Tobacco, and Barley Leaves

	Enzyme activity (μmol h^{-1} mg^{-1} Chl)			
Plant	Ascorbate peroxidase	Monodehydroascorbate reductase	Dehydroascorbate reductase	Glutathione reductase
Pea	77 ± 15	6.8 ± 1.9	36 ± 0.5	83 ± 20
Maize	77 ± 17	7.4 ± 0.7	38 ± 7.1	30 ± 0.8
Tobacco	170 ± 17	—	30 ± 1	58 ± 28
Barley	77 ± 7	28 ± 2.3	100 ± 7	57 ± 5

TABLE 2
The Effects of Methyl Viologen Treatment (5 μ*M*) on the Ascorbate and Dehydroascorbate Contents of Normal and Transformed Tobacco Leaves in the Light

Leaf type	Treatment	Ascorbate content[a]	Dehydro-ascorbate content[a]	Total pool
Untransformated tobacco leaf	Control	1190	131	1321
	+5 μ*M* methyl viologen	798	431	1229
Tobacco leaf expressing bacterial glutathione reductase in the cytosol	Control	1064	119	1263
	+5 μ*M* methyl viologen	850	281	1131

[a] nmol mg^{-1} Chl.

proportion of these enzymes is localized in the chloroplasts,[34] it is clear that sufficient activity is also present in the cytosol to drive the H_2O_2-scavenging system in green and nongreen tissues.[39,46] Indeed, when leaves from transformed tobacco plants expressing high levels of bacterial glutathione reductase activity in the cytosol were treated with the herbicide, methyl viologen,[60] the ascorbate pool was better protected against oxidation than in the untransformed controls, even though the site of action of methyl viologen is the thylakoid membrane of the chloroplast (Table 2). These transgenic plants also exhibited a lower susceptibility to methyl viologen in terms of the extent of visible foliar damage, but, interestingly, they were no more resistant to ozone than were the controls.[61] These observations suggest that the increased cytosolic glutathione reductase activity increased the rate of supply of electrons to oxidized ascorbate in the cytosol in this situation of oxidative stress, and that the ascorbate pool in the chloroplast was consequently afforded some added protection,[60] presumably because of cycling of ascorbate between the cytosolic and chloroplast compartments (Figure 3). These experiments demonstrate that the cytosolic reactions of the ascorbate-glutathione cycle can help to support the chloroplast ascorbate-glutathione cycle in the stroma, possibly by the transport system across the chloroplast envelope.[37,38]

The operation of the ascorbate-glutathione cycle provides effective protection against the constant stress of aerobiosis. In addition, it has become clear that one of the plant responses to environmental stress that often increases the potential for oxidant formation is to modify the activities of the antioxidant enzymes. The activity of ascorbate peroxidase and also of other enzymes of the cycle has been reported to be increased in leaf tissues in response to the number of stress conditions, for example drought;[62] high light intensity;[48] acclimation to excess light;[63] fumigation with ozone;[64] and treatment with low levels of the herbicide, methyl viologen.[65] This increase in the activities of antioxidant enzymes appears to be critical for tolerance to a broad away of stresses. Furthermore, tolerance to one type of oxidant stress can confer resistance to other stresses. For example, biotypes of the weed *Conyza borariensis* have developed resistance to the herbicide paraquat (methyl viologen). Resistance is associated with high constitutive levels of plastid antioxidant enzymes.[66] These biotypes also show cross-resistance to high irradiance and SO_2 pollution.[65,67]

Resistance appears to be governed by a single pleiotropic gene that coordinates stress tolerance.[68] Reported increases in activity are not large (two- to threefold), but they may critically affect stress tolerance. However, the overall increase in total activity of the enzymes may be less important than the induction of isoenzymes better adapted to the new conditions with the cell imposed by the stress.

C. ASCORBATE OXIDASE AND OTHER ENZYMES

In this discussion we have considered the benefits of keeping the ascorbate pool in its reduced form to protect metabolism against oxidative stress. However, there may be certain conditions where controlled oxidation of ascorbate may be of use, and thus oxidation of ascorbate is required. It must be remembered that H_2O_2 has important functions in plant cells, such as the lignification of cell walls, the metabolism of indole-3-acetic acid, and defenses against attack by pathogens.[69,70] H_2O_2 has a role in elicitor signal transduction. As part of the elicitor system, it is involved in the early stages of the hypersensitive reaction that results in the production of phytoalexins in the cell of plants.[70] It thus, has an important role in the plant-defense systems.[70] For this reason, ascorbate oxidation and ascorbate oxidases may be necessary. In addition, oxidation of ascorbate may provide a redox system in which NADPH is oxidized via the ascorbate-glutathione cycle (Figure 2).

Ascorbate oxidases are copper-containing enzymes containing 8 to 12 atoms of copper per hologenzyme.[71,72] They catalyze the oxidation of ascorbate using molecular oxygen,[73,74] as shown in Reaction 6:

$$2 \text{ ascorbate} + O_2 + 2H^+ \longrightarrow \text{dehydroascorbate} + H_2O \qquad (6)$$

Note that only the ascorbate monoanion serves as the substrate, and that the oxidase has an absolute requirement for molecular O_2 and produces only water, its pH optimum being pH 5.6.

Ascorbate oxidases have been studied intensively, but their functional role has always remained a matter of speculation. The copper-containing ascorbate oxidase occurs only in plants, and is essentially a different enzyme from the one found in animals. Ascorbate oxidases have been purified from fruit and vegetables[71-76] where they occur at high levels in association with the cell wall and cytoplasm.[32,33] Ascorbate oxidase was one of the first proteins to be used in the study of the effects of phytochrome on enzyme activity. Exposure to short periods of red light increases enzyme activity; for example, in mustard seedlings, exposure to 36 h red light increased ascorbate oxidase activity sevenfold.[77] However, the concentration of ascorbate oxidase protein did not change significantly, but rather posttranslational activation by phytochrome was suggested.[77] The ubiquitous distribution of ascorbate oxidase may suggest that it also fulfills a role in the balance between reduction and oxidation in plant cells, as part of the ascorbate system, but favoring ascorbate oxidation rather than maintaining the ascorbate pool.

It is important to note that the existence of enzymes that can phosphorylate ascorbic acid to ascorbic acid 2-phosphate has been reported in the liver[78] and in bacteria such as *Pseudomonas azotocolligans* KY 4661.[79] These "ascorbic acid phosphorylating enzymes" appear to be either phosphatases or phosphotransferases.[79] Whether such reactions can occur in plants remains to be demonstrated.

IV. FUNCTION

The metabolic requirement for oxygen creates an apparent paradox for all aerobic cells.[11,80] Aerobic metabolism is highly efficient, but the O_2 molecule is predisposed to a univalent pathway of reduction.[80] This is unfortunate because it generates highly reactive derivatives of oxygen. The major endogenous sources of these oxidants are the mitochondrial and photosynthetic electron transport chains.[11,80,81] Similarly, photodynamic events leading to the generation of singlet oxygen[80] pose a continuous threat to the photosynthetic membranes. The major function of ascorbate in plants is to combat these deleterious events, and thus prevent metabolic disruption and cellular damage.

A. PROTECTION AGAINST ENZYME INACTIVATION

The presence of the ascorbate-dependent scavenging system is essential for the continued function of photosynthesis in the light. Several of the enzymes of the reductive pentose phosphate pathway are activated in the light by thiol modulation, via the thioredoxin system, and these are extremely sensitive to oxidation.[10,82,83] The reduction state of these enzymes, which determines the activation state, reflects the balance between the flux of reducing equivalents through the electron transport chain (that causes their activation) and the oxidizing environment of the stroma that continuously favors inactivation.[84,85] H_2O_2 interferes with the delicate balance of this system, since it favors oxidation and prevents thioredoxin-dependent reductive activation of these enzymes.[10,82,83] Because of this, H_2O_2 is a potent inhibitor of photosynthetic CO_2 assimilation. The principal site of inhibition by H_2O_2 is that of the light-modulated enzymes of the Benson-Calvin cycle.[10,83] Five enzymes of the

Benson-Calvin cycle require activation upon illumination. Of these, fructose-1,6-bisphosphatase, sedoheptulose-1,7-bisphosphatase, phosphoribulokinase, and NADP-dependent glyceraldehyde phosphate dehydrogenase are activated by the thioredoxin system.[84,85] This light-activation system is readily reversible; the natural terminal oxidant upon darkening is molecular O_2.[85] The reduced form of purified fructose-1,6-bisphosphatase has been shown to be inactivated by H_2O_2.[82] H_2O_2-dependent inhibition of CO_2 assimilation is associated with increased [ATP] to [ADP] and [Triosephosphate] to [3-phosphoglycerate] ratios. Large increases in sedoheptulose-1,7-bisphosphate occur in the presence of H_2O_2, and fructose-1,6-bisphosphate also accumulates.[10] This evidence shows that sedoheptulose-1,7-bisphosphatase and fructose-1,6-bisphosphatase are principal targets for H_2O_2 inactivation *in vivo*.[10,82] Pentose monophosphates also accumulate, suggesting the inhibition of phosphoribulokinase, as well as the two bisphosphatases. The formation of 6-phosphogluconate is also greatly stimulated.[10] Hence, glucose 6-phosphate dehydrogenase that is normally inactivated by reduction upon illumination becomes oxidized and activated. The presence of 6-phosphogluconate may serve to further inhibit CO_2 assimilation by competing for the active site of ribulose-1,5-bisphosphate carboxylase. Because H_2O_2 inhibits photosynthesis, even at low concentrations (10 μM), efficient and rapid removal of H_2O_2 is required in order to maintain maximal rates of photosynthesis. In addition, H_2O_2 will oxidize all susceptible thiol groups, and many other enzymes are vulnerable to H_2O_2-dependent inactivation; for example, H_2O_2 slowly inactivates chloroplast superoxide dismutase.

B. PROTECTION AGAINST THE EFFECTS OF EXCESS IRRADIANCE

The irradiance to which a plant is exposed provides the energy to drive photosynthesis, and also provides information concerning the light environment in terms of quantity and quality that enables the plant to make metabolic adjustments in the short-term, and adaptations in structure in the long-term to optimize the photosynthetic processes in line with the prevailing light environment (Figure 8). When irradiance is low, the absorption and use of light energy is highly efficient. Nearly all of the absorbed light energy is used for photosynthesis (Figure 9). In this situation, the rate of photosynthesis is limited by the availability of light energy. Under these conditions, maximum quantum yields are observed because of regulation that allows photosystems to be excited at appropriate rates. This balance is achieved, at least in part, by reversible phosphorylation of a mobile population of the light-harvesting complexes that adjusts the light-energy absorption of the photosystems. Thus, at low irradiances, very little of the absorbed light energy is dissipated, since most, if not all, can be used effectively for photosynthesis (Figure 9). However, at light levels approaching saturating (Figure 9A) and beyond, the rate of photosynthesis is limited by both the availability for CO_2 and the metabolic capacity for CO_2 assimilation and sucrose synthesis. At light saturation, more light is absorbed than can be used effectively for photosynthesis, and the excess must be removed by effective dissipative processes (Figure 9B). These result in decreases in the measured quantum efficiencies of the photosystems.[86] Such regulated losses in efficiency are necessary in order to maintain optimum redox poise and prevent over-reduction and

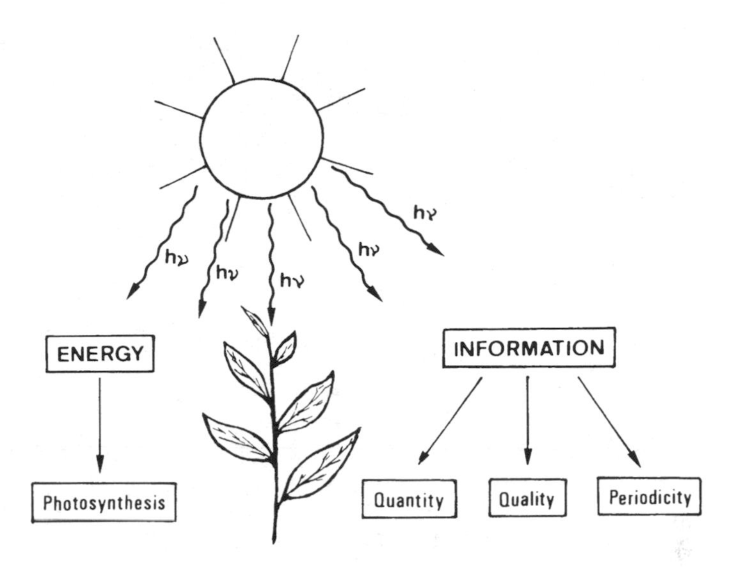

FIGURE 8. Diagrammatic representation of the function of light in plant responses to the environment.

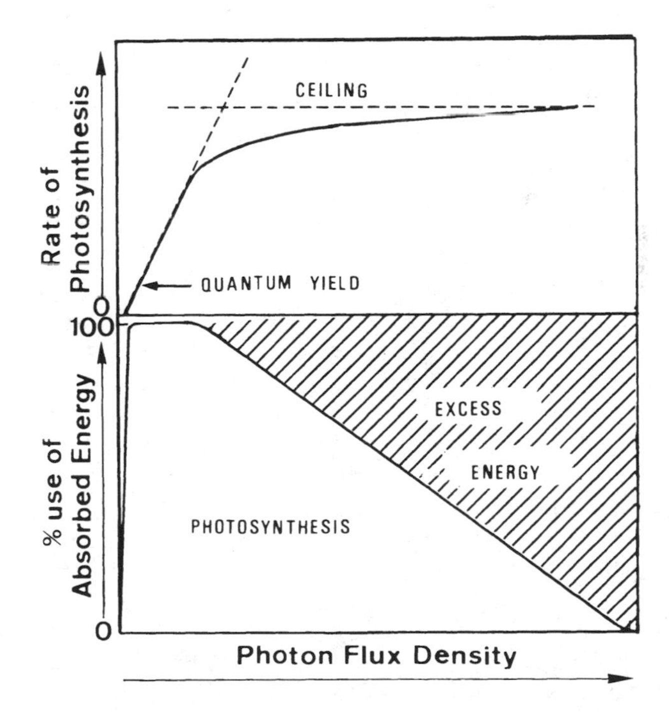

FIGURE 9. (A) A theoretical light-response curve for photosynthesis in leaves. (B) A diagrammatic representation of the relative amount of excitation energy used in photosynthesis as irradiance (photon flux density) is increased.

photodestruction. Prolonged exposure to light levels where excitation energy exceeds the capacity for carbon assimilation can lead to the light-induced loss of photosynthetic capacity, called photoinhibition.[86] In its role as an antioxidant, ascorbate is a component of the regulatory system that serves to coordinate supply and demand in photosynthesis.[86] The rate of electron transport is coordinated with the rate at which the products of electron flow (ATP and NADPH) can be effectively used by metabolism.[86] However, imbalances between the energy supply and the capacity of metabolism to make effective use of the absorbed energy do occur. In this situation, the effect of excess irradiance is the generation of excess reducing power and over-reduction of the photosynthetic electron transport intermediates. This unavoidably leads to the stimulation of pseudocyclic electron flow to oxygen, and results in increased formation of O_2^- and H_2O_2 (Figure 7). The ascorbate-glutathione cycle removes these free radicals using reducing power (NADPH) provided by the electron transport chain (Figures 2 and 7). Surplus reducing power can thus be effectively lost in a process illustrated simply in Reaction 5. Furthermore, ascorbate is intimately involved with the generation of the processes that lead to the thermal dissipation of excess excitation energy, as discussed below.

1. Pseudocyclic Electron Flow

Oxygen has several functions in photosynthesis. It has a protective role, since it is involved in the processes that avoid over-reduction of the stroma and electron transport chain. Excess energy, in terms of reducing power, can be lost through the processes of photorespiration and pseudocyclic electron flow. In the latter process, oxygen serves as an electron acceptor and produces superoxide (Figure 7). The capacity for oxygen reduction may be higher than 20 μmol h^{-1} mg^{-1} chlorophyll.[87,88] Electron transport to oxygen poises the electron carriers in terms of preventing over-reduction, and is a coupled process leading to ATP synthesis.[87,88] In addition, it is possible that pseudocyclic electron flow enables the electron transport system to produce an ATP-$2e^-$ ratio of 1:5. This is required for optimal photosynthetic CO_2 assimilation. Noncyclic electron flow alone (Figure 7) is believed to provide an ATP-$2e^-$ ratio of only 1:33. Thus, ATP synthesis driven by pseudocyclic electron flow could make up the balance by producing ATP, but not NADPH.

Pseudocyclic electron flow results in the formation of H_2O_2. The H_2O_2-scavenging system is highly efficient, and H_2O_2 is a good electron acceptor.[58,59] The production of H_2O_2 as a result of pseudocyclic electron flow and its subsequent scavenging could act as a valve reaction preventing over-reduction and photoinhibitory damage.[59] It provides a means of eliminating excess reducing power in the chloroplast.

Oxygen competes with NADP for reducing equivalents (Figure 7). Electron flow to oxygen has always been considered to be rather low under steady-state conditions, accounting for less than 20% of the net steady-state rate of oxygen evolution during CO_2 assimilation at high irradiances.[87,88] However, when CO_2 assimilation is restricted and the NADP pool becomes relatively reduced, electron flow to O_2 is favored. Such a situation arises during the induction phase of photosynthesis that follows a transition from darkness to light or a transition from low irradiance to high irradiance.[89-92] In this situation, over-reduction is necessary to force the thiol-dependent activation of the light-modulated enzymes of the Benson-Calvin cycle.[90] During the induction phase, the processes of psudocyclic and cyclic electron flow serve to generate ATP because noncyclic electron flow is limited by the availability of NADP.[90] Once the thiol-modulated enzymes of the Benson-Calvin cycle are activated, the turnover of this cycle increases the demand for NADPH, and ATP also increases such that the NADPH to NADP ratio falls to a level very close to the dark value, and noncyclic electron flow is predominant.[90]

2. The Generation of Transthylakoid ΔpH and Its Associated Fluorescence-Quenching Components

Pseudocyclic electron flow is clearly important in the induction phase of photosynthesis.[89-92] For this reason, it has been suggested that pseudocyclic electron flow coupled to H_2O_2-dependent O_2 evolution is the basis of the complex induction kinetics of chlorophyll, a fluorescence known as the "Kautsky effect".[59,93-96] When chlorophyll a fluorescence measurements are made on leaves illuminated following a period of darkness, the yield of chlorophyll fluorescence displays characteristic changes in intensity that accompany the induction phase of photosynthesis.[95] The onset of actinic illumination causes a rapid, biphasic rise in fluorescence yield from the dark level (Fo) to an intermediate level (I), then rising to a maximum peak level (P). From P, the fluorescence decays to a stationary (S) level, then rises to a second maximum (M), before moving to a terminal (T) level.[93-95] Anoxia considerably slows down the progress of these transitions.[94] Oxygen is thus required for the processes that lead to the changes in chlorophyll a fluorescence quenching before it reaches its steady-state (T) level. The characteristic changes in chlorophyll a fluorescence quenching may, therefore, be intimately associated with pseudocyclic electron flow coupled to the H_2O_2-scavenging ascorbate-glutathione cycle.

Electron transport generates the proton motive force that drives ATP synthesis. The proton motive force across the thylakoid membrane consists almost entirely of ΔpH because of counter-ion movement which negates most of the contribution from the electrical potential difference. This ΔpH is generated by proton release into the thylakoid lumen and accompanying alkalization of the chloroplast stroma. High ΔpH has direct effects on the quantum efficiency of photosystem II (PS II). In the short-term, the quantum yield of the photochemistry of PS II is regulated by the nonphotochemical quenching of excitation energy induced by a high transthylakoid ΔpH.[86,95,97] This major regulatory response allows harmless dissipation of excess excitation energy as heat in the chlorophyll antenna of PS II, and is of fundamental importance in the protection of the photosynthetic processes against photodestruction.[97] The generation of a high ΔpH triggers a rapidly reversible loss of quantum efficiency in PS II essentially by increasing the rate constant of thermal energy dissipation with respect to that of photochemistry. Much of the nonphotochemical quenching of chlorophyll fluorescence is linked to the formation of the transthylakoid ΔpH, and is thus called energy-dependent quenching, or qE. The mechanism of nonphotochemical quenching related to the generation of ΔpH, caused by thylakoid membrane energization, is not yet fully understood.[97] However, it is clear that the addition of H_2O_2 to isolated intact chloroplasts causes pronounced fluorescence quenching due to changes in both photochemical quenching (caused by charge separation at the PS II reaction center) and nonphotochemical quenching (nonradiative deexcitation) processes.[59] Furthermore, these changes in chlorophyll a fluorescence quenching are eliminated if electron transport is inhibited by 3-(3,4-dichloro-phenyl)-1,1-dimethylurea (DCMU).[93] Clearly, metabolism of H_2O_2 leads to membrane energization and formation of a transthylakoid ΔpH, and hence, nonphotochemical quenching of chlorophyll a fluorescence. When H_2O_2 is generated within the chloroplast as a result of pseudocyclic electron flow and the H_2O_2 is metabolized by the ascorbate-glutathione cycle, the ΔpH which is formed as a result of these processes favors the operation of nonradiative dissipation of excitation energy (and the associated nonphotochemical quenching of chlorophyll a fluorescence).

H_2O_2 scavenging may be responsible for most of the nonphotochemical quenching of chlorophyll a fluorescence that develops when CO_2 assimilation is limited, for example, during the induction phase of photosynthesis. This would lead to a decrease in the efficiency of PS II activity.[59,94,96] In this way, the ascorbate system is involved not only as an antioxidant protecting against H_2O_2 and other oxygen radicals, but it is also intimately associated with the processes of photosynthetic control and energy dissipation.[86,94,96] Ascorbate may have a further role in the regulation of the quantum efficiency of PS II as described below.

C. MEMBRANE EFFECTS

1. Formation of Zeaxanthin

Ascorbate is the co-factor for the thylakoid enzyme violaxanthin depoxidase which forms zeaxanthin (Reaction 7). Demmig-Adams and Adams (Chapter 4) have shown that the formation of zeaxanthin is strongly correlated with the appearance of certain processes of energy dissipation in the thylakoid membrane. They suggest that zeaxanthin has a direct role in nonradiative energy dissipation, an increase in the rate constant for nonradioactive dissipation being correlated with the appearance of this xanthophyll pigment in the thylakoid membranes. The formation of the nonphotochemical quenching component, qE does not require the presence of zeaxanthin, but the presence of zeaxanthin amplifies qE formation such that it can be formed in the absence of a high ΔpH.[97] The presence of zeaxanthin depends on the regulation a complex reaction sequence known as the xanthophyll or violaxanthin cycle.[21,51] This cycle consists of the light-dependent de-epoxidation of violaxanthin, via antheraxanthin, to zeaxanthin, as shown in Reaction 7, and the light-independent epoxidation of zeaxanthin via antheraxanthin to violaxanthin, as shown in Reaction 8.

$$\text{violaxanthin} \xrightarrow{\text{2H ascorbate}} \text{antheraxanthin} \xrightarrow{\text{2H ascorbate}} \text{zeaxanthin} \quad (7)$$

$$\text{zeaxanthin} \xrightarrow{\text{NADPH} + \text{H}^+ + \text{O}_2} \text{antheraxanthin} \xrightarrow{\text{NADPH} + \text{H}^+ + \text{O}_2} \text{violaxanthin} \quad (8)$$

The violaxanthin de-epoxidase is located within the thylakoid membrane, and has an acidic pH optimum such that it is active only in the light when the trans-thylakoid ΔpH is adequate.[21,51] Ascorbate has a significant function in the generation of the membrane-bound quencher, zeaxanthin, which is produced only under stress conditions when the dissipation of excess light energy in the thylakoid membranes is necessitated by unfavorable conditions (see Demmig-Adams and Adams, Chapter 4).

Ascorbate thus participates in the defense mechanisms of the plant, both indirectly in the avoidance mechanisms (such as zeaxanthin formation), and directly in the detoxification processes by removal of oxygen radicals. The size and reduction state of the ascorbate pool reflect the status of the leaf in terms of its antioxidant capacity, and effects on the ascorbate pool may provide an early approximation of the effects of stress on the photosynthetic apparatus.

If excess excitation energy exceeds the capacity of the pathways of energy dissipation, then photodamage to the PS II reaction center occurs. Plants grown in conditions of low light are far more susceptible to this photoinhibition than leaves of plants grown in high light. The levels of ascorbate and other active oxygen scavengers are lower in plants grown in low light or shade conditions.[34,48] Shade-grown plants have less of the xanthophyll cycle components than do plants grown in high light (see Demmig-Adams and Adams, Chapter 4). Such a decreased capacity in the protective systems results in increased susceptibility to active oxygen species and free radicals. Thus, to some degree, the growth conditions affect the capacity of plant to withstand high light stress. However, the immediate environmental conditions to which leaves are exposed are most important in determining the susceptibility to light-induced damage. Even with plants acclimated to high light, exposure to environmental stress (which effectively lowers the ceiling of maximum photosynthesis, Figure 9) increases the susceptibility to light-induced damage and free-radical production. At low temperatures, for example, zeaxanthin synthesis is much reduced or totally inhibited (see Demmig-Adams and Adams, Chapter 4). Also, the ascorbate pool can become extensively oxidized and subsequently decreased in size.[98] In chilling-sensitive cucumber leaves, α-tocopherol was the primary antioxidant affected by a chill at high light, but the ascorbate pool also rapidly became considerably oxidized.[98] Chilling-sensitive maize genotypes have been found to have less of the enzymes involved in oxygen scavenging than chilling-resistant genotypes.[99] Even in the presence of the H_2O_2-scavenging systems, a threefold increase in the detected concentration of H_2O_2 was found as a result of a cold treatment in the dark in winter wheat seedlings.[100] Seedlings incubated at 28°C in the dark showed decreased levels of H_2O_2[100] compared to those kept at low temperatures.

2. Recycling of α-Tocopherol

Ascorbate is required for the regeneration of α-tocopherol (Reaction 3). This antioxidant is incorporated into the photosynthetic membranes, and serves to reduce the possibility of damage caused by singlet oxygen and lipid peroxides. α-Tocopherol (vitamin E) is well known as an efficient lipophilic antioxidant. It traps lipid peroxy radicals in the deep inner regions of membranes, and in the process is oxidized to the relatively stable tocopheroxyl radical. This is re-reduced back to tocopherol by ascorbic acid. In plant cells, α-tocopherol is located mainly in the chloroplast, and is present at high concentrations both in the chloroplast envelope and in the thylakoid membranes. α-Tocopherol is an essential constitutive component of the membrane. Ascorbate and tocopherol have the same charge as their reduced compounds at physiological pH. Tocopherol, like ascorbate, may function via the transfer of single hydrogen atoms, in order to minimize the possibility of reduction of molecular oxygen to superoxide while retaining its radical-scavenging abilities.[26] The ascorbate pool represents a reservoir of antioxidant potential which is needed to regenerate α-tocopherol.

3. Effects on Electron Transport Components

It is also pertinent to ask whether ascorbate has any effect on the redox state of any of the carriers of the electron transport chain. The general consensus is that the reaction of ascorbate with components such as cytochrome f is too slow to compete with their reduction by PS II. Ascorbate may act as a hydrogen donor, rather than an electron donor, to the electron transport system, since it must be noted that at physiological pH, ascorbate is a poor electron donor, but a good donor of single hydrogen atoms.[26] Ascorbate is capable of forming ionic bonds with many biological molecules at physiological pH. Binding to receptor molecules does not directly affect ascorbate, but brings about localized alterations of the redox potential, and may accelerate charge transfer reactions.[101] This binding will affect the steric molecular configuration and electrical charge of macromolecules or specialized regions of membranes.

Ascorbate will also influence the reactivity of the molecular complex to which it is bound because the highly active hydroxy group of the C_2 is not directly affected by the monovalent intermolecular bond at carbon atom 3.[101] Ascorbate binds to the thylakoid membranes in a ratio of 0.5 ± 0.1 mg ascorbate mg^{-1} chlorophyll.[102] The binding of ascorbate to stroma-exposed components of the electron transport chain will have some significance for their function. Transfer of electrons from ascorbate to several components of the electron transport chain, e.g., plastoquinone, plastocyanin, cytochrome b_{559}, and cytochrome f (Figure 10) has been demonstrated. Although such transport of electrons is not sufficiently rapid to compete with the light-driven transport of electrons from PS II, it might affect the poising of the redox state of the electron carriers.

V. SUMMARY

Ascorbate affects many diverse reactions and physiological processes in plants. It is probable that in most, if not all, the role of ascorbate is as a reductant protecting, and occasionally participating in, many metabolic activities. The ascorbate pool is a reservoir of reducing power. Ascorbic acid is an exceptional antioxidant that scavenges, either directly or indirectly, all of the damaging free radicals commonly encountered in plant cells. Most importantly, ascorbate plays a pivotal role in the

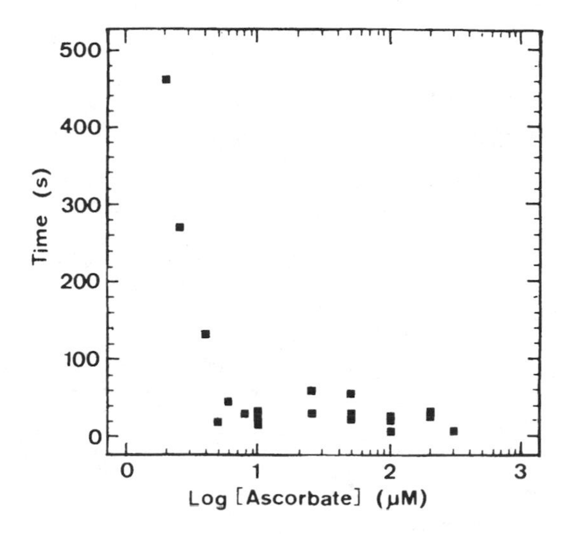

FIGURE 10. The effect of ascorbate concentration on the rate of dark re-reduction of cytochrome f by ascorbate, in isolated spinach thylakoids in the presence of the inhibitor, DCMU, and the electron acceptor, methyl viologen.

scavenging of H_2O_2. Plants have evolved specific ascorbate peroxidases that fulfill functions analogous to the glutathione peroxidases in animal tissues. Ascorbate protects the thiol groups of enzymes and proteins from oxidation caused by the constant stress of aerobiosis. The most sensitive of these are the thiol-modulated enzymes of the photosynthetic carbon reduction cycle. In addition to its role as a primary antioxidant, ascorbic acid is also an important secondary antioxidant. It maintains the α-tocopherol pool which scavenges radicals in the deep inner regions of membranes. Hydrophilic ascorbate and hydrophobic tocopherol may be considered to be rather unique in terms of their high capacities for free-radical scavenging combined with very low reactivities with oxygen, i.e., they do not readily donate electrons to oxygen. In contrast to the relatively poor understanding of the precise role of ascorbate in many of the biochemical and physiological functions of plants, precise roles for ascorbate and the ascorbate-dependent H_2O_2-scavenging system are evident in photosynthesis. The ascorbate system is not only important as a detoxification mechanism, but it also has an obligatory role in the regulation of electron flow *in vivo*. The reduction of O_2 to H_2O_2 in photosynthesis and its subsequent metabolism to H_2O, via the ascorbate-glutathione cycle, does not require ATP, but conversely aids membrane energization and ATP synthesis. In addition, the contribution of this system to membrane energization affords protection, since it triggers processes in the membrane that lower the intrinsic quantum yield of PS II thus helping to prevent photoinhibitory damage. Furthermore, ascorbate is an essential co-factor for the synthesis of zeaxanthin, which also makes an important additional contribution to the lowering of the efficiency of photochemistry in PS II under stress conditions.

REFERENCES

1. **Padh, H.,** Cellular functions of ascorbic acid, *Biochem. Cell Biol. (Biochimie Biologie Cellulaire)*, 68, 1166, 1990.
2. **Chinoy, J. J.,** The role of ascorbic acid in growth, differentiation and metabolism of plants, Chonoy, N. J., Ed., Nijhoff/Junk, The Hague, Netherlands, 1984, 1.
3. **Loewus, F. A.,** Ascorbic acid and its metabolic products, *The Biochemistry of Plants*, Vol. 14, Academic Press, New York, 1988, 85.
4. **Giroud, A.,** *L'Acide Ascorbique dans la Cellule et les Tissus*, Borntrager, Berlin, 1938, 1.
5. **Rabinowitch, E. I.,** Ascorbic acid in green plants. I. Ascorbic acid in the chloroplasts, in *Photosynthesis*, Vol. 1, 1945, 269.
6. **Foyer, C. H. and Halliwell, B.,** Presence of glutathione and glutathione reductase in chloroplasts: a proposed role in ascorbic acid metabolism, *Planta*, 133, 21, 1976.
7. **Groden, D. and Beck, E.,** H_2O_2 destruction by ascorbate-dependent systems from chloroplasts, *Biochim. Biophys. Acta*, 546, 426, 1979.
8. **Nakano, Y. and Asada, K.,** Spinach chloroplasts scavenge hydrogen peroxide on illumination, *Plant Cell Physiol.*, 21, 1295, 1980.
9. **Nakano, Y. and Asada, K.,** Hydrogen peroxide is scavenged by ascorbate-specific peroxidase in spinach chloroplasts, *Plant Cell Physiol.*, 22, 867, 1981.
10. **Kaiser, W. M.,** Reversible inhibition of the Calvin cycle and activation of oxidative pentose phosphate cycle in isolated chloroplasts by hydrogen peroxide, *Planta*, 145, 377, 1979.
11. **Asada, K. and Takahashi, M.,** Production and scavenging of active oxygen in photosynthesis, in *Photoinhibition*, Kyle, D. J., Osmond, C. B., and Arntzen, C. J., Eds., Elsevier, Amsterdam, 1987, 227.
12. **Volk, S. and Feierabend, J.,** Photoinactivation of catalase at low temperature and its relevance to photosynthetic and peroxide metabolism in leaves, *Plant, Cell and Environ.*, 12, 701, 1989.
13. **Overbaugh, J. M. and Fall, R.,** Characterization of a selenium-independent glutathione peroxidase from *Euglena gracilis*, *Plant Physiol.*, 77, 437, 1985.
14. **Criqui, M. C., Jamet, E., Parmentier, Y., Marbach, J., Durr, A., and Fleck, J.,** Isolation and characterisation of a plant cDNA showing homology to animal glutathione peroxidases, *Plant Mol. Biol.*, 18, 623, 1992.
15. **Jablonski, P. P. and Anderson, J. W.,** Role of a flavonoid in the peroxide dependent oxidation of glutathione catalyzed by pea extracts, *Phytochemistry*, 23, 1865, 1984.
16. **Anbar, M. and Neta, P.,** A compilation of specific bimolecular rate constants for the reactions of hydrated electrons, hydrogen atoms and hydroxyl radicals with inorganic and organic compounds in aqueous solution, *Int. J. Appl. Radiat. Isotopes*, 18, 493, 1967.
17. **Nishikimi, M.,** Oxidation of ascorbic acid with superoxide anion generated by the xanthine-xanthine oxidase system, *Biochim. Biophys. Res. Commun.*, 63, 463, 1975.
18. **Nakano, Y. and Asada, K.,** Purification of ascorbate peroxidase in chloroplasts; its inactivation in ascorbate-depleted medium and reactivation by monodehydroascorbate radical, *Plant Cell Physiol.*, 28, 131, 1987.
19. **Tanaka, K., Takenchi, E., Kubo, A., Sakaki, T., Haraguchi, K., and Kawamura, Y.,** Two immunologically different isozymes of ascorbate peroxidase from spinach leaves, *Arch. Biochem. Biophys.*, 286, 371, 1991.
20. **Finckh, B. F. and Kunert, K. J.,** Vitamins C and E: an antioxidative system against herbicide-induced plants, *J. Agric. Food Chem.*, 33, 574, 1985.
21. **Hager, A.,** Die reversiblen, lichtabhängigen xanthophyllumwandlungen im chloroplasten, *Ber Deutsch. Bot. es. Bd.*, 88, S27, 1975.
22. **Foyer, C., Rowell, J., and Walker, D.,** Measurement of the ascorbate content of spinach leaf protoplasts and chloroplasts during illumination, *Planta*, 157, 239, 1983.
23. **Law, M. Y., Charles, S. A., and Halliwell, B.,** Glutathione and ascorbic acid in spinach (*Spinacia oleracea*) chloroplasts. The effect of hydrogen peroxide and paraquat, *Biochem. J.*, 210, 899, 1983.

24. **Foyer, C. H., Dujardyn, M., and Lemoine, Y.,** Responses of photosynthesis and the xanthophyll and ascorbate-glutathione cycle to changes in irradiance, photoinhibition and recovery, *Plant Physiol. Biochem.*, 27, 751, 1989.
25. **Stegmann, H. B., Schuler, P., Ruff, H. J., Knollmüller, M., and Loreth, W.,** Ascorbic acid as an indicator of damage to forest. A correlation with air quality, *Z. Naturforsch.*, 46C, 67, 1991.
26. **Njus, D. and Kelly, P. M.,** Vitamins C and E donate single hydrogen atoms *in vivo, FEBS Lett.*, 284, 147, 1991.
27. **Hossain, H. A., Nakano, Y., and Asada, K.,** Monodehydroascorbate reductase in spinach chloroplasts and its participation in regeneration of ascorbate for scavenging of hydrogen peroxide, *Plant Cell Physiol.*, 25, 385, 1984.
28. **Hossain, H. A. and Asada, K.,** Purification of dehydroascorbate reductase from spinach and its characterisation as a thiol enzyme, *Plant Cell Physiol.*, 25, 85, 1984.
29. **Jablonski, P. P. and Anderson, J. W.,** Light-dependent reduction of dehydroascorbate by ruptured pea chloroplasts, *Plant Physiol.*, 67, 1239, 1981.
30. **Foyer, C. H., Lelandais, M., Edward, E. A., and Mullineaux, P. M.,** The role of ascorbate in plants; interactions with photosynthesis and regulatory significance, in *Active Oxygen/Oxydative Stress and Plant Metabolism,* Vol. 6, Pell, E. and Steffen, K., Eds., 131, 1991.
31. **Polle, A., Chakrabarti, K., Schürmann, W., and Rennenberg, H.,** Composition and properties of hydrogen peroxide decomposing systems in extracellular and total extracts from needles of Norway spruce (*Picea abies* L., karst), *Plant Physiol.*, 94, 312, 1990.
32. **Hallaway, M., Phethean, P. D., and Taggart, J.,** A critical study of the intracellular distribution of ascorbate oxidase and a comparison of the kinetics of the soluble and cell wall enzyme, *Phytochemistry,* 9, 935, 1970.
33. **Chichiricco, G., Ceru, M. P., D'Alessandro, A., Oratore, A., and Avigliano, L.,** Immunohistochemical localisation of ascorbate oxidase in *Cucurbita pepo* medullosa, *Plant Sci.*, 64, 61, 1989.
34. **Gillham, D. J. and Dodge, A. D.,** Hydrogen-peroxide scavenging systems within pea chloroplasts. A quantitative study, *Planta,* 167, 246, 1986.
35. **Franke, V. W. and Heber, U.,** Uber die quantitative Verteilung der Ascorbinsäure innerhalb der Pflanzenzelle, *Z. Naturforsch. Teil,* B19, 1146, 1964.
36. **Rennenberg, H.,** Glutathione metabolism and possible biological roles in higher plants, *Phytochemistry,* 21, 2771, 1982.
37. **Anderson, J. W., Foyer, C. H., and Walker, D. A.,** Light-dependent reduction of dehydroascorbate and uptake of exogenous ascorbate by spinach chloroplasts, *Planta,* 158, 442, 1983
38. **Beck, A., Burkert, A., and Hofmann, M.,** Uptake of L-ascorbate by intact spinach chloroplasts, *Plant Physiol.*, 73, 41, 1983.
39. **Dalton, D. A., Russel, S. A., Hanus, F. J., Pascoe, G. A., and Evans, H. J.,** Enzymatic reactions of ascorbate and glutathione that prevent peroxide damage in soybean root nodules, *Proc. Natl. Acad. Sci. U.S.A.*, 83, 3811, 1986.
40. **Klapheck, S., Zimmer, I., and Crosse, H.,** Scavenging of hydrogen peroxide in the endosperm of *Ricinus communis* by ascorbate peroxidase, *Plant Cell Physiol.*, 31, 1005, 1990.
41. **Chen, G. X. and Asada, K.,** Ascorbate peroxidase in tea leaves: occurrence of two isozymes and the differences in their enzymatic and molecular properties, *Plant Cell Physiol.*, 30, 987, 1989.
42. **Mittler, R. and Zilinskas, A. A.,** Purification and characterisation of pea cytosolic ascorbate peroxidase, *Plant Physiol.*, 97, 962, 1991.
43. **Edwards, E. A., Rawsthorne, S., and Mullineaux, P. M.,** Subcellular distribution of multiple forms of glutathione reductase in leaves of pea *(Pisum sativum* L.), *Planta,* 180, 278, 1990.
44. **Hossain, M. A. and Asada, K.,** Monodehydroascorbate reductase from cucumber is a flavin adenine dinucleotide enzyme, *J. Biol. Chem.*, 260, 12920, 1985.
45. **Ito, A., Hayashi, A., and Yoshida, T.,** Participation of a cytochrome 6-like hemoprotein of the outer mitochondrial membrane (OM cytochrome 6) in NADH-semi dehydroascorbic acid reductase activity of rat liver, *Biochem. Biophys. Res. Commun.*, 101, 591, 1981.

46. **Dalton, D. A., Post, C. J., and Langeberg, L.,** Effects of ambient oxygen and of fixed nitrogen on concentrations of glutathione, ascorbate and associated enzymes in soybean root nodules, *Plant Physiol.*, 96, 812, 1991.
47. **Jablonski, P. P. and Anderson, J. W.,** Light-dependent reduction of hydrogen peroxide by ruptured pea chloroplasts, *Plant Physiol.*, 69, 1407, 1982.
48. **Gillham, D. J. and Dodge, A. D.,** Chloroplast superoxide and hydrogen peroxide scavenging systems from pea leaves: seasonal variations, *Plant Sci.*, 50, 105, 1987.
49. **Schupp, R. and Rennenberg, H.,** Diurnal changes in the glutathione content of spruce needles (*Picea abies* L.), *Plant Sci.*, 57, 113, 1988.
50. **Esterbauer, H., Grill, D., and Welt, R.,** The annual rhythm of the ascorbate acid system in needles of *Picea abies*, *Z. Pflanzenphysiol.*, 98, 393, 1980.
51. **Yamamoto, H. Y.,** Biochemistry of the violaxanthin cycle in higher plants, *Pure Appl. Chem.*, 51, 639, 1979.
52. **Shigeoka, S., Nakano, Y., and Kitaoka, S.,** The biosynthetic pathway of L-ascorbic acid in *Euglena gracilis*, *Z. J. Nutr. Sci. Vitaminol.*, 25, 299, 1979.
53. **Loewus, F. A.,** Tracer studies on ascorbic acid formation in plants, *Phytochemistry*, 2, 109, 1963.
54. **Saito, K., Nick, J. A., and Loewus, F. A.,** D-glucosone and L-sorbosone, purative intermediates of L-ascorbic acid biosynthesis in detached bean and spinach leaves, *Plant Physiol.*, 94, 1496, 1990.
55. **Loewus, M. W., Bedgar, D. L., Saito, K., and Loewus, F. A.,** Conversion of L-sorbosone to L-ascorbic acid by a NADP-dependent dehydrogenase in bean and spinach leaves, *Plant Physiol.*, 94, 1492, 1990.
56. **Grün, M. and Loewus, F. A.,** L-ascorbic acid biosynthesis in the euryhaline diatom *Cyclotella cryptica*, *Planta*, 160, 6, 1984.
57. **Helsper, J. P. F. G., Kagan, L., Hilby, C. L., Maynard, T. M., and Loewus, F. A.,** L-ascorbic acid biosynthesis in *Ochromonas danica*, *Plant Physiol.*, 69, 465, 1981.
58. **Anderson, J. W., Foyer, C. H., and Walker, D. A.,** Light-dependent reduction of hydrogen peroxide by intact spinach chloroplasts, *Biochim. Biophys. Acta*, 724, 69, 1983.
59. **Schreiber, U., Reising, H., and Neubauer, C.,** Contrasting pH optima of light-driven O_2 and H_2O_2 reduction in spinach chloroplasts as measured via chlorophyll fluorescence quenching, *Z. Naturforsch.*, 46C, 635, 1991.
60. **Foyer, C. H., Lelandais, M., Galap, C., and Kunert, K. J.,** Effects of elevated glutathione reductase activity on the cellular glutathione pool and photosynthesis in leaves under normal and stress conditions, *Plant Physiol.*, 97, 863, 1991.
61. **Aono, M., Kubo, A., Saji, H., Natori, T., Tanaka, K., and Kondo, N.,** Resistance to active oxygen toxicity of transgenic *Nicotiana tabacum* that expresses the gene for glutathione reductase in *Escherichia coli*, *Plant Cell Physiol.*, 32, 691, 1991.
62. **Smirnoff, N. and Colombe, S. V.,** Drought influences the activity of enzymes of the chloroplast hydrogen peroxide scavenging system, *J. Exp. Bot.*, 39, 1097, 1988.
63. **Schoner, S. and Krause, G. H.,** Protective systems against active oxygen species in spinach: response to cold acclimation in excess light, *Planta*, 180, 383, 1990.
64. **Mehlhorn, H., Cottam, D. A., Lucas, P. W., and Wellburn, A. R.,** Induction of ascorbate peroxidase and glutathione reductase activities by interactions of mixtures of air pollutants, *Free Rad. Res. Comm.*, 3, 193, 1987.
65. **Shaaltiel, Y., Glazer, A., Bocion, P. F., and Gressel, J.,** Cross tolerance to herbicidal and environmental oxidants of plant biotypes tolerant to paraquat, sulfur dioxide and ozone, *Pest. Biochem. Physiol.*, 31, 13, 1988.
66. **Shaaltier, Y. and Gressel, J.,** Multienzyme oxygen radical detoxifying system correlated with paraquat resistance in *Conyza bonariensis*, *Pest. Biochem. Physiol.*, 25, 22, 1986.
67. **Jansen, M. A. K., Malan, C., Shaaltiel, Y., and Gressel, J.,** Mode of photooxidant resistance to herbicides and xenobiotics, *Z. Naturforsch.*, 45C, 463, 1990.
68. **Shaaltier, Y., Chua, N. H., Gepstein, S., and Gressel, J.,** Dominant pleiotropy controls enzymes co-segregating with paraquat resistance in *Conyza bonariensis*, *Theor. Appl. Genet.*, 75, 850, 1988.

69. **Takahama, U. and Egashiva, T.,** Peroxidases in vacuoles of *Vicia faba* leaves, *Phytochemistry,* 30, 73, 1991.
70. **Apostol, I., Heinstein, P. F., and Low, P. S.,** Rapid stimulation of an oxidative burst during elicitation of cultured plant cells, *Plant Physiol.,* 90, 109, 1989.
71. **Lee, M. H. and Dawson, C. R.,** Ascorbate oxidase: further studies on the purification of the enzyme, *J. Biol. Chem.,* 248, 6596, 1973.
72. **Marchesini, A. and Kronek, P. M. H.,** Ascorbate oxidase from *Cucurbita pepo medullosa.* New method of purification and re-investigation of properties, *Eur. J. Biochem.,* 101, 65, 1979.
73. **Butt, V. S.,** Direct oxidases and related enzymes, in *The Biochemistry of Plants. A Comprehensive Treatise. Metabolism and Respiration,* Vol. 2, Davies, D. D., Ed., Academic Press, New York, 1980, 85.
74. **Malmström, B. G., Andreasson, L. E., and Reinhammer, B.,** Copper-containing oxidases and superoxidase dismutase, in *The Enzymes,* Vol. 12B, 3rd Ed., Boyer, P. D., Ed., Academic Press, New York, 1975, 507.
75. **Avigliano, L., Vecchini, P., Sirianni, P., Marcozzi, G., Marchesini, A., and Mondovi, B.,** A reinvestigation on the quaternary structure of ascorbate oxidase from *Cucurbiat pepo medullosa, Mol. Cell. Biochem.,* 56, 107, 1983.
76. **Nakamura, T., Nakino, N., and Ogura, Y.,** Purification and properties of ascorbate oxidase from cucumber, *J. Biochem.,* (Tokyo), 64, 189, 1968.
77. **Leaper, L. and Newbury, H. J.,** Phytochrome control of the accumulation and rate of synthesis of ascorbate oxidase in mustard cotyledons, *Plant Sci.,* 64, 79, 1989.
78. **Miyosaki, T., Sato, M., Yoshinaka, R., and Sakaguchi, M.,** Synthesis of ascorbyl 2-P by a liver enzyme of rainbow trout, *Comp. Biochem. Physiol.,* B100, 4, 1991.
79. **Koizumis, S., Maruyama, A., and Fujio, T.,** Purification and characterisation of ascorbic acid phosphorylating enzyme from *Pseudomonas azotocolligans, Agric. Biol. Chem.,* 54, 3235, 1990.
80. **Rabinowitch, H. D. and Fridovich, I.,** Superoxide radicals, superoxide dismutases and oxygen toxicity in plants, *Photochem. Photobiol.,* 37, 679, 1983.
81. **Rich, P. R. and Bonner, W. D., Jr.,** The sites of superoxide generation in higher plant mitochondria, *Arch. Biochem. Biophys.,* 188, 206, 1978.
82. **Charles, S. A. and Halliwell, B.,** Effect of hydrogen peroxide on spinach (*Spinacia oleracia*) chloroplast fructose bisphosphatase, *Biochem. J.,* 189, 373, 1980.
83. **Tanaka, K., Otsubo, T., and Kondo, N.,** Participation of hydrogen peroxide in the inactivation of Calvin cycle SH enzymes in SO_2-fumigated spinach leaves, *Plant Cell Physiol.,* 23, 1009, 1982.
84. **Cseke, C. and Buchanan, B. B.,** Regulation of the formation and utilisation of photosynthate in leaves, *Biochim. Biophys. Acta,* 853, 43, 1986.
85. **Leegood, R. C.,** Enzymes of the Calvin cycle, in *Methods in Plant Biochemistry,* Vol. 3, Lea, P. J., Ed., Academic Press, London, 1990, 15.
86. **Foyer, C., Furbank, R., Harbinson, J., and Horton, P.,** The mechanisms contributing to photosynthetic control of electron transport by carbon assimilation in leaves, *Photosynth. Res.,* 25, 83, 1990.
87. **Egneus, H., Heber, U., Mathiesen, U., and Kirk, M.,** Reduction of oxygen by the electron transport chain of chloroplasts during assimilation of carbon dioxide, *Biochim. Biophys. Acta,* 408, 252, 1975.
88. **Heber, U., Egneus, U., Hanck, M., Jensen, M., and Köster, S.,** Regulation of photosynthetic electrontransport and photophosphorylation in intact chloroplasts and leaves of *Spinacia oleracea, Planta,* 143, 41, 1978.
89. **Marsho, T. V., Behrens, P. N., and Radmer, K. J.,** Photosynthetic oxygen reduction in isolated intact chloroplasts and cells from spinach, *Plant Physiol.,* 64, 656, 1979.
90. **Foyer, C. H., Lelandais, M., and Harbinson, J.,** Control of the quantum efficiencies of PSI and PSII, electron flow and enzyme activation following dark to light transitions in pea leaves. The relationship between NADP/NADPH ratios and NADP-malate dehydrogenase activation state, *Plant Physiol.,* 99, 979, 1992.

91. **Furbank, R. T., Badger, M. R., and Osmond, C. B.,** Photoreduction of oxygen in mesophyll chloroplasts of C4 plants. A model system for studying an *in vivo* Mehler reaction, *Plant Physiol.*, 73, 1038, 1983.
92. **Avelange, M. H. and Rébeillé, F.,** Mass spectrometric determination of O_2 gas exchange during a dark-to-light transition in higher plant cells, *Planta,* 183, 158, 1991.
93. **Neubauer, C. and Schreiber, U.,** Photochemical and non-photochemical quenching of chlorophyll fluorescence induced by hydrogen peroxide, *Z. Naturforsch.*, 44C, 262, 1989.
94. **Neubauer, C. and Schreiber, U.,** Dithionite-induced fluorescence quenching does not reflect reductive activation in spinach chloroplasts, *Bot. Acta,* 102, 314, 1989.
95. **Krause, G. H. and Weis, E.,** Chlorophyll fluorescence and photosynthesis; the basics, *Annu. Rev. Plant Physiol. Plant Mol. Biol.*, 42, 313, 1991.
96. **Schreiber, U. and Neubauer, C.,** O_2 dependent electron flow, membrane energisation and the mechanism of non-photochemical quenching of chlorophyll fluorescence, *Photosynth. Res.*, 25, 279, 1990.
97. **Horton, P., Ruban, A. V., Pascal, A. A., Noctor, G., and Young, A. J.,** Control of the light-harvesting function of chloroplast membranes by aggregation of the LCHII chlorophyll-protein complex, *FEBS Lett.*, 292, 1, 1991.
98. **Wise, R. R. and Naylor, A. W.,** Chilling-enhanced photooxidation. Evidence for the role of singlet oxygen and superoxide in the breakdown of pigments and endogenous antioxidants, *Plant Physiol.*, 83, 278, 1987.
99. **Jahnke, L. S., Hull, M. R., and Long, S. P.,** Chilling stress and oxygen metabolism enzymes in *Zea mays* and *Zea diploperennis, Plant, Cell Environ.*, 14, 97, 1991.
100. **Okuda, T., Matsuda, Y., Yamanaker, A., and Sagiska, S.,** Abrupt increase in the level of hydrogen peroxide in leaves of winter wheat is caused by cold treatment, *Plant Physiol.*, 97, 1265, 1991.
101. **Bensch, K. G.,** On a possible mechanism of action of ascorbic acid: formation of ionic bonds with biological molecules, *Biochem. Biophys. Res. Commun.*, 101, 312, 1981.
102. **Aristarkov, A. I., Nikandrov, V. V., and Krasnovikii, A. A.,** Ascorbate permeability of chloroplast thylakoid membrane: reduction of plastoquinones and cytochrome f. Translated from *Biokhimiya,* 52, 2051, in *Biochemistry,* 52, 1776, 1987.

Chapter 3

CAROTENOIDS

Kenneth E. Pallett and Andrew J. Young

TABLE OF CONTENTS

0-8493-6328-4/93/$0.00 + $.50

I. INTRODUCTION

Carotenoids are C_{40} isoprenoids or tetraterpenes which are present in all green tissues where they are exclusively located in the chloroplast. Carotenoids are also responsible for the yellow to red pigmentation of many plant tissues (i.e., roots, flowers, fruit) where they are located in other plastids such as chromoplasts. There are two classes of carotenoids in plants: (1) the carotenes which are hydrocarbon compounds and (2) the xanthophylls which contain one or more oxygen functions.

Within the chloroplast, carotenoids have important functional roles, particularly as accessory light-harvesting pigments and as photoprotective agents. Carotenoids have long been recognized as important antioxidants both *in vivo* and *in vitro*. Chloroplasts in photosynthetic organisms are capable of generating a number of potentially toxic oxygen species such as singlet oxygen ($^1O_2$), superoxide (O_2^-), hydrogen peroxide, and hydroxyl radicals. The production of these reactive species is a consequence of normal chloroplastidic processes, and is generally dealt with by several protective agents, i.e., superoxide dismutase, ascorbate, as well as carotenoids. Under stress conditions (e.g., low temperature, enhanced photoinhibition, or herbicide treatment), normal protective processes may become overloaded, and cellular destruction, including pigment degradation and lipid peroxidation, will result.

This chapter aims to examine the antioxidant role of carotenoids, specifically the carotenes in photosynthetic tissues of higher plants, and will provide examples of the conditions that will result in oxidative degradation of these pigments. In Chapter 4, the role of the xanthophyll cycle will be reviewed. In order to put the functions of carotenoids into context, their biosynthesis will be reviewed, together with a consideration of the compartmentalization and regulation of biosynthesis and their distribution within the chloroplast.

II. BIOSYNTHESIS

The biosynthesis of carotenoids can be conveniently divided into five stages, namely: (1) the general isoprenoid pathway, (2) the formation of phytoene, (3) the desaturation of phytoene, (4) cyclization, and (5) the introduction of oxygen functions (Figure 1).

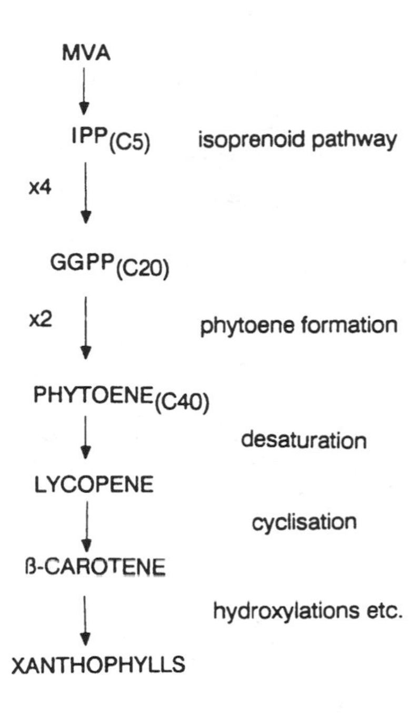

FIGURE 1. An overview of carotenoid biosynthesis.

A. THE GENERAL ISOPRENOID PATHWAY

This pathway involves the formation and subsequent conversion of the isoprenoid precursor, mevalonic acid (MVA) to isopentenyl pyrophosphate (IPP). This common five-carbon unit of isoprenoids undergoes isomerization to dimethyl allyl pyrophosphate (DMAPP), which then condenses with another IPP to form geranyl pyrophosphate (GPP). Condensation with two further IPP units results in the formation of geranyl geranyl pyrophosphate (GGPP), which is the precursor for carotenoids. This polymerization of IPP occurs in the stroma of the chloroplast. For a more detailed consideration of the general isoprenoid pathway, its enzymology and significance in carotenoid biosynthesis, see Qureshi and Porter,[1] Poulter and Rilling,[2] Dogbo and Camara,[3] and Lutzow and Beyer.[4]

B. THE FORMATION OF PHYTOENE

The condensation of two GGPP molecules to form phytoene is the first step unique to carotenoid biosynthesis. Figure 2 shows the mechanism of this condensation, which proceeds via a C_{40} intermediate prephytoene pyrophosphate from which a proton is stereospecifically lost to give rise to the (15-Z) or 15,15*cis* isomer of phytoene in higher plants.[5]

This reaction is catalyzed by phytoene synthetase, which is a soluble enzyme located in the stroma of chloroplasts and other plastids.[6] Its exact location is uncertain, but it is likely to be peripherally attached to an inner envelope and/or thylakoid membranes, and may be easily dissociated.[7,8] Phytoene synthetase has

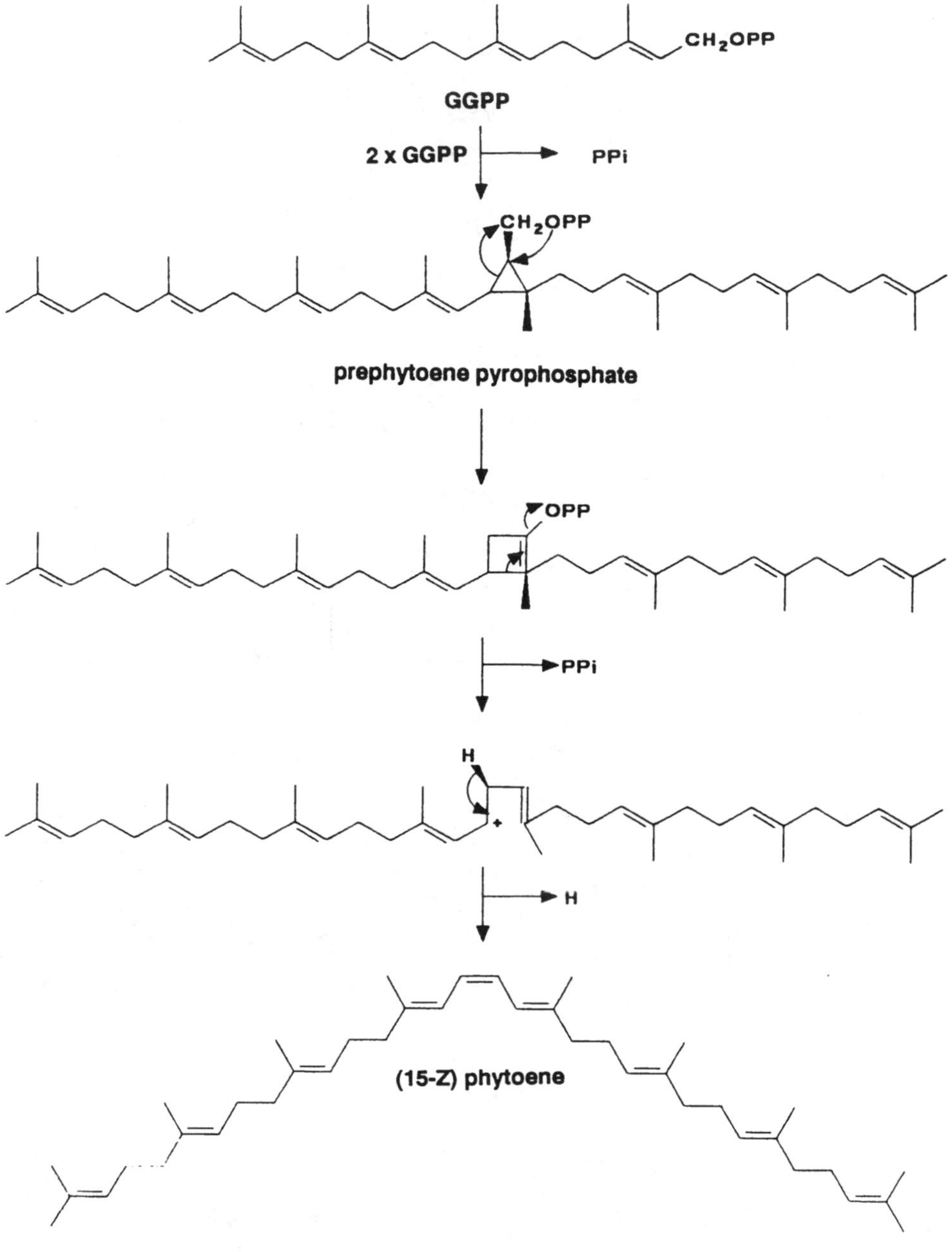

CH_2OPP

GGPP

2 x GGPP → PPi

CH_2OPP

prephytoene pyrophosphate

OPP

→ PPi

H

+

→ H

(15-Z) phytoene

FIGURE 2. The mechanism of (15-*Z*) phytoene biosynthesis.

been purified from *Capsicum annum* chromoplasts, and it has a strict requirement for Mn^{2+}, but no other co-factors appear necessary for phytoene synthesis.[9]

C. THE DESATURATION OF PHYTOENE

Phytoene (7,8,11,12,7′,8′,11′,12′-octahydro-ψ-ψ-carotene) undergoes a series of desaturation reactions to form phytofluene, ζ-carotene, neurosporene, and finally lycopene (ψ,ψ-carotene) (Figure 3). At each stage, two hydrogen atoms are lost by transelimination.[5,7] An alternative to the symmetrical ζ-carotene is the formation of its asymmetrical isomer (7,8,11,12ψ,ψ-carotene), which, upon desaturation, yields neurosporene (Figure 3).

An isomerization from 15-Z(*cis*) to all-E (*trans*) occurs during the desaturation sequence. The evidence from higher plant studies indicates that this isomerization occurs at the level of phytofluene, as shown in Figure 3, although this reaction may not be the same for all organisms.[7,8] The reaction mechanism of desaturation is generally believed to be a simple dehydrogenation; however, it is possible that a two-stage hydroxylation followed by dehydration may occur.[7,10]

The desaturation of phytoene is catalyzed by phytoene desaturase (dehydrogenase), which is membrane bound to the inner envelope and/or thylakoid. Its exact location is uncertain. Desaturase activity was reported in the envelope of isolated spinach chloroplasts.[11] However, in chloroplast fractions from *Raphanus sativus*, desaturase activity was detected only in the thylakoids.[12] In isolated chromoplasts, desaturase activity was reported to be on the inner envelope membrane.[13,14]

The enzyme has yet to be isolated and purified from higher plants, and many questions remain to be answered regarding phytoene desaturation. *In vitro* studies with isolated plastids indicate that two desaturase enzymes are involved in lycopene biosynthesis, phytoene desaturase which converts 15-Z phytoene to all-E ζ-carotene, and ζ-carotene desaturase then forms all-E lycopene via neurosporene (Figure 3).[8] However, the four desaturation stages may be catalyzed by a single protein with different requirements for co-factors, O_2 and light for each stage.[15,16] Oxygen appears to be a requirement for desaturation, and light is necessary for isomerization in isolated daffodil chromoplasts.[16,17] Evidence appears to support an oxidoreductase acting as a redox mediator between the desaturase and O_2.[16]

D. CYCLIZATION

Once the carotene end group has reached the stage of lycopene desaturation, cyclization can occur to give six-membered β- or ϵ-rings, at one or both ends of the carotene molecule (Figure 4). Initial proton attack at C2 leads to an intermediate carbonium ion which undergoes proton loss at either C4 or C6 to form ϵ- or β-rings, respectively, the common rings of major higher plant carotenoids (Figure 5).[5,7,8]

Lycopene cyclase is membrane bound and closely associated with the desaturase. It has yet to be isolated and purified, and similar uncertainty to that for desaturase(s) exists on its location. Cyclization has been reported in the thylakoid fraction of *R. sativus* chloroplasts[12] and in the membrane fraction, presumably

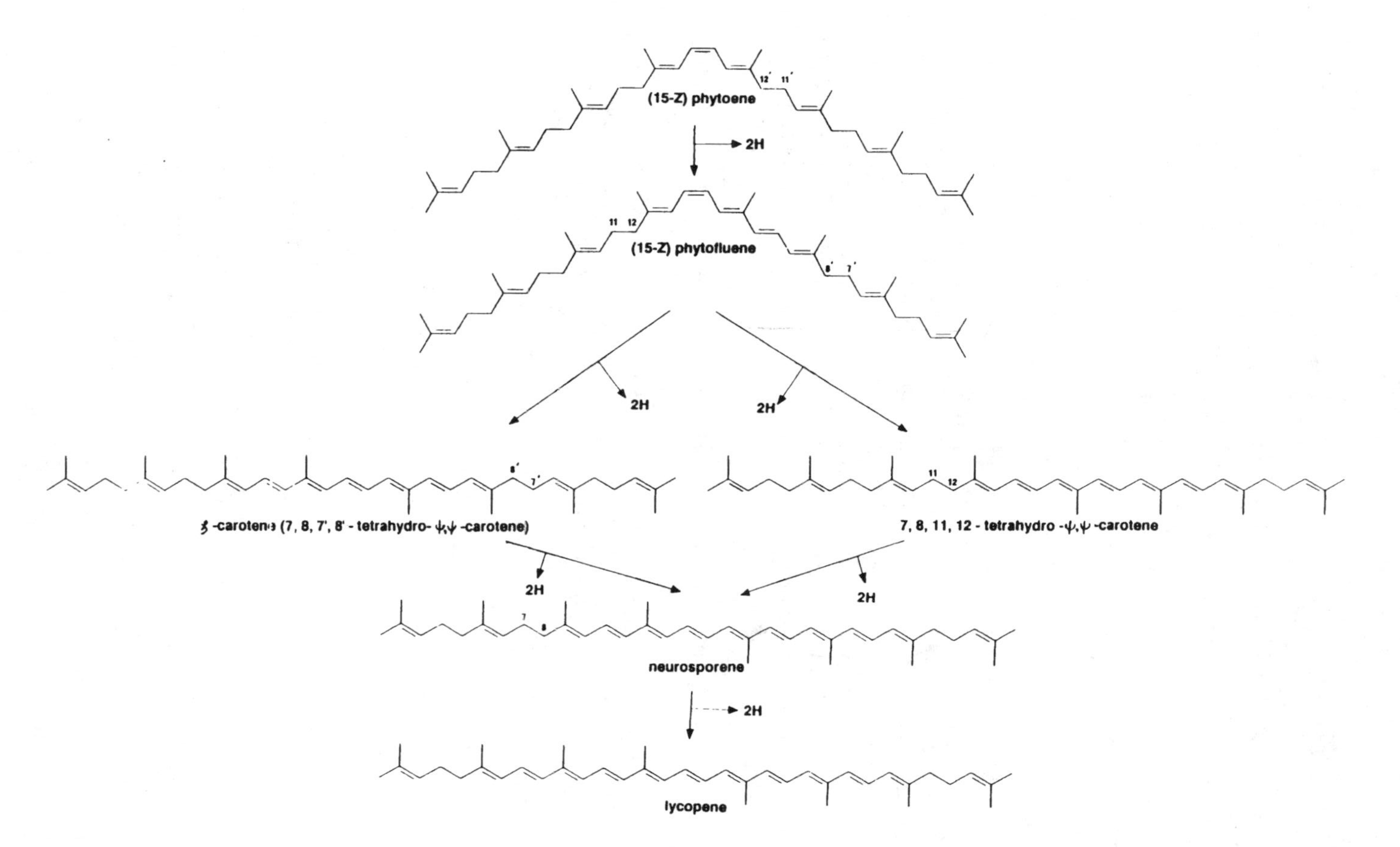

FIGURE 3. The sequence of desaturations and isomerization from (15-*Z*) phytoene to lycopene.

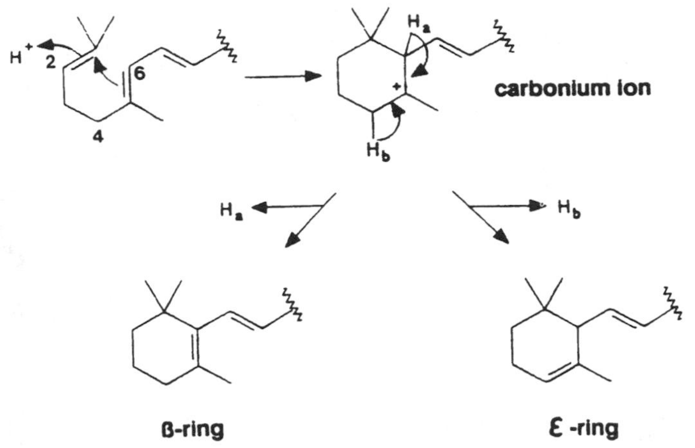

FIGURE 4. The mechanism of cyclization to form β- and ε-rings.

envelope of *C. annum* and *Narcissus pseudonarcissus* chromoplasts.[17-19] Although desaturase activity was detected in spinach chloroplast envelopes, cyclase activity was absent.[11] In the *N. pseudonarcissus* chromoplast membranes, oxygen was shown to be necessary for desaturation; however, its absence was necessary for cyclization.[17] Hence, care should be taken with *in vitro* studies to ensure anaerobic conditions for lycopene cyclase, since aerobic conditions may have led to the reported absence of cyclase with *R. sativus*.[12] If this is indeed the situation, then there will be a strict membrane microenvironment for aerobic desaturation and anaerobic cyclization in the plastid.

E. INTRODUCTION OF OXYGEN FUNCTIONS

The hydroxylation and epoxidation reactions leading to the xanthophylls are believed to occur at the postcyclization stage.[7,8] Again, little is known about the enzymes and their location; however, the involvement of mixed function oxidases with a putative cytochrome P450 is assumed.[7,8] There is a close association between desaturase(s) and cyclase, as xanthophyll biosynthesis has been reported in the thylakoid fraction from *R. sativus* chloroplasts,[12] and in the envelope membrane fraction from *C. annum* chromoplasts.[19]

The xanthophyll or violaxanthin cycle, the function which will be covered by Demmig-Adams and Adams in Chapter 4, involves the light-dependent de-epoxidation of violaxanthin to zeaxanthin via antheraxanthin. This conversion is catalyzed by a light-dependent de-epoxidase. An epoxidase catalyzes the reverse reaction that requires O_2 and NADPH, and occurs under limited light conditions.[20] The location of the de-epoxidase and epoxidase is uncertain, with evidence suggesting envelope and thylakoid membrane as sites for the xanthophyll cycle. The biosynthetic pathway from carotenes to violaxanthin is proposed to occur via zeaxanthin, which then undergoes epoxidation.[20]

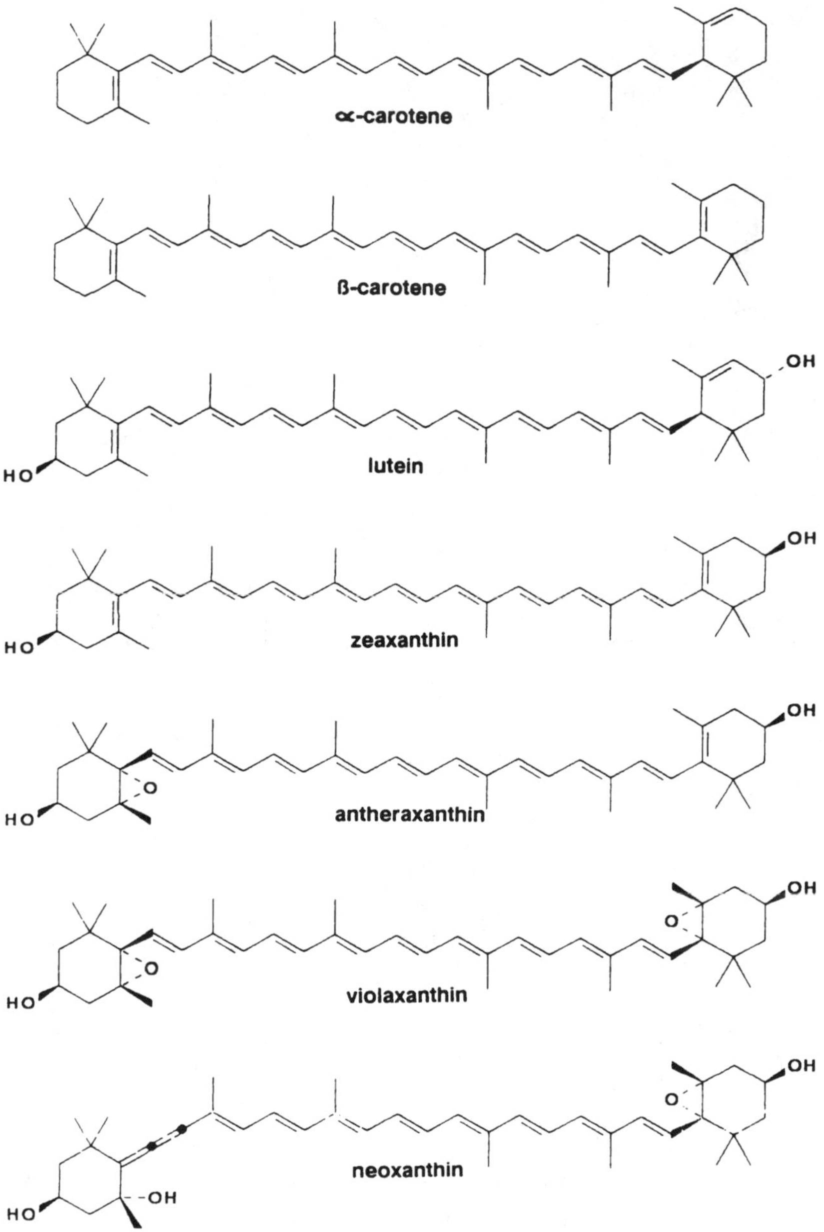

α-carotene

β-carotene

lutein

zeaxanthin

antheraxanthin

violaxanthin

neoxanthin

FIGURE 5. The structures of the major chloroplast carotenoids.

III. LOCATION OF CAROTENOIDS IN THE CHLOROPLAST

Before the review of the photoprotective function of carotenoids, a brief consideration will be given to the distribution of carotenoids within the chloroplast and also to factors which regulate their biosynthesis and distribution.

A. DISTRIBUTION OF CAROTENOIDS

Carotenoid distribution in the thylakoid membrane is now well characterized.[22] They are essential components of the pigment protein complexes, namely the reaction- or core-center of photosystem I (CC I) reaction- or core-center of photosystem II (CC II) and their respective light-harvesting or antenna complexes (LHCs I and II).[21,22]

β-carotene appears to be the only carotenoid component of CC I. It is also the major, if not the only, carotenoid associated with the pigment-protein complexes of CC II. Lutein may be associated with CC II.[21,22] LHC I contains lutein, neoxanthin, and violaxanthin, plus β-carotene, whereas the three xanthophylls are the major components of LHC II complexes.[21] β-carotene may be present in LHC II, but only as a minor component.[22]

There is some disagreement regarding other plastid locations for carotenoids. The xanthophylls, particularly violaxanthin, have been reported to be present in the chloroplast envelope.[23] However, an absence or very low levels of carotenoids has also been reported in the envelope,[24] but, more recently, high performance liquid chromatography (HPLC) analysis of purified envelopes clearly reveals carotenoids present in the envelope fraction (Table 1). Moreover, the distribution of carotenoids differs from that of the thylakoid fraction, with 90% comprising xanthophylls, two thirds of which is violaxanthin (Table 1), and, in accordance with previous reports,[23] 26% of thylakoid carotenoid is β-carotene due to its location in the core centers of PS I and PS II. Lutein is the major xanthophyll, almost twice that of violaxanthin, reflecting the higher proportion of lutein in the LHC's in the thylakoid.[21] The high proportion of violaxanthin in the envelope fraction reflects its formation from zeaxanthin in this fraction.[20]

B. REGULATION OF THE BIOSYNTHESIS AND DISTRIBUTION OF CAROTENOIDS IN THE CHLOROPLAST

Light is the main regulator for chloroplast development, and therefore for carotenoid biosynthesis.[5] Dark grown plants are etiolated and devoid of chloroplasts and chlorophyll, and possess limited carotenoids, principally xanthophylls. However, it is only upon illumination that normal chloroplast development with associated bulk pigments (carotenoids and chlorophylls) occurs. The reaction- or core-center complexes, including their carotenoids (principally β-carotene), are synthesized first, and subsequently, LHCs with the xanthophylls are synthesized.[5] In normal plant development, chloroplasts develop directly from proplastids in dividing meristematic cells.

TABLE 1
The Carotenoid Content of Thylakoid and Envelope Fractions of Isolated Spinach Chloroplasts

	Thylakoid μg/mg protein	Envelope μg/mg protein	Envelope μg/mg protein (corrected)
β-Carotene	8.7 (26.1)	1.3 (10.3)	0.9 (8.0)
Violaxanthin	6.9 (20.7)	7.4 (58.7)	7.1 (63.4)
Antheraxanthin	0.6 (1.8)	0.4 (3.2)	0.3 (2.7)
Lutein	12.8 (38.4)	2.8 (22.2)	2.3 (20.5)
Neoxanthin	4.3 (12.9)	0.7 (5.6)	0.6 (5.4)
Total carotenoid	33.3 (100)	12.6 (100)	11.2 (100)
Chlorophylls	151.8	5.7	—

Note: Figures in parentheses represent % total carotenoid. Thylakoid and envelope fractions were obtained from isolated intact spinach chloroplasts by the methods previously described.[25] Carotenoid analysis was carried out as previously described.[26] The chlorophyll content of the envelope membrane is due to contamination of this fraction by thylakoid membrane, and the corrected data for envelope takes into account thylakoid pigment concentration.

From Pallett, K. E. and Joyard, J., unpublished data.

Chloroplast carotenoid composition is greatly influenced by light, as exhibited by sun- and shade-adapted plants. In the latter, there is a greater xanthophyll proportion of chloroplast carotenoids due to a greater proportion of LHCs.[27,28]

Carotenoid biosynthesis is closely integrated with chlorophyll biosynthesis and the development of the photosynthetic apparatus. At least three phases of carotenoid synthesis are evident:[5]

1. Bulk synthesis — development of photosynthetic apparatus
2. Turnover — in photosynthetically active chloroplasts (maintenance of normal carotenoid composition)
3. Synthesis induced by environmental conditions

It seems likely that each of these is regulated differently, although, as stated above, light is of major importance, which is also central to the function of carotenoids.[5] During chloroplast development the biosynthesis of carotenoids, chlorophyll, pigment-proteins, and thylakoid membranes are very closely integrated, occurring simultaneously. Evidence for this arises from studies with carotenoid-inhibiting

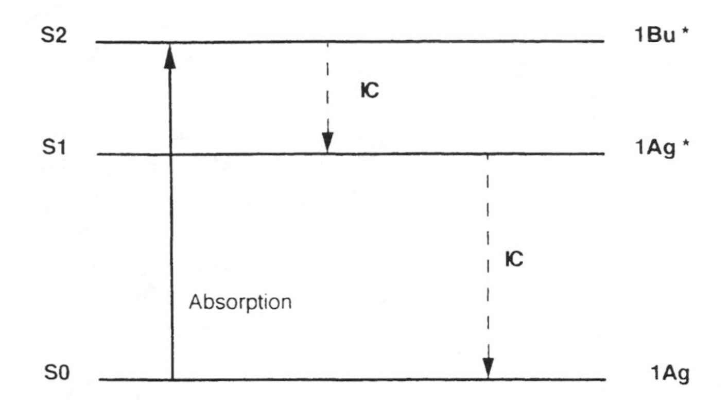

FIGURE 6. Light absorption deactivation scheme for carotenoids. IC, intersystem crossing; S_0, ground state; S_1, 1st excited state; S_2, 2nd excited state. (Redrawn from Truscott, T. G., *J. Photochem. Photobiol. B*, 6, 359, 1990. With permission.)

herbicides such as diflufenican, a specific inhibitor of the desaturation of phytoene to phytofluene.[29] Leaf tissue treated with diflufenican develops devoid of carotenoids and chlorophylls, and plastids appear in the proplastid stage with little evidence of internal thylakoid membranes.[29]

IV. FUNCTION

A. GENERAL PHOTOCHEMISTRY

The behavior of carotenoids in photosynthetic systems (light-harvesting, quenching of triplet state sensitizers, and antioxidative function) is largely governed by their photochemical properties. The ground state (S_0) of carotenoid molecules is termed the ^1Ag state. The absorption of light will promote this to an excited state (S_2), termed $^1Bu^*$. A further low-lying symmetry-forbidden excited singlet state (S_1) also exists, and is termed $^1Ag^*$ (Figure 6). Recent experimental evidence has shown that the excited $^1Bu^*$ state decays very rapidly by fast internal conversion to the $^1Ag^*$ state, and it is this that is involved in energy transfer processes in photosynthetic systems. The C_{40} carotenoids typical of those found in higher plants and in many algae have recently been shown to be weakly fluorescent.[30,31] It is these properties of the carotenoids found in all photosynthetic systems that make them efficient in energy-transfer processes, necessary for light harvesting and photoprotection. Readers are referred to more detailed reviews of the photochemistry of carotenoids.[22,32,33]

B. LIGHT HARVESTING

The role of carotenoids in light-harvesting processes in photosynthetic organisms has recently been extensively reviewed.[21,22,32,34-36]

The bulk of the chloroplast carotenoids are associated with the light-harvesting complexes in the thylakoid membrane. The carotenoids present in these antenna complexes, the bulk of which are mainly xanthophylls, have the ability to absorb light in the wavelength range of 400 to 500 nm. Energy is transferred from the carotenoid-excited singlet state ($^1Car^*$) to S_0 chlorophyll (^1Chl) by singlet-singlet energy transfer (Reactions 1 to 3). This energy-transfer process is generally recognized to proceed with very high efficiency, itself suggesting a close proximity of carotenoid and chlorophyll molecules. *In vitro* studies using synthetic caroteno-porphyrin molecules have shown that high efficiencies can be achieved provided that a precise structural arrangement is maintained between the carotenoid and chlorophyll molecules.[37] The possible role of carotenoid radical species in electron transfer processes in light harvesting has been discussed for a number of years, although there is still a lack of clear evidence for their involvement in anything but artificial systems.[33]

$$^1Car + h\nu \longrightarrow {}^1Car^* \quad (1)$$

$$^1Car^* \ (Bu, S_2) \longrightarrow {}^1Car^* \ (2Ag, S_1) + heat \quad (2)$$

$$^1Car^* + {}^1Chl \longrightarrow {}^1Car + {}^1Chl^* \quad (3)$$

This light-harvesting ability of carotenoids allows for a particularly efficient use of light in the blue region. The carotenoids in *Lactuca sativa* (which account for less than 25% of total pigment) are responsible for nearly half of the absorbance of light in the 390 to 520 nm range.[38] The main function of the carotenoids in these complexes should be the harvesting of light energy, although it is clear that they also have a photoprotective role.

C. PHOTOPROTECTION

While the light harvesting carried out by the carotenoids of the antenna complexes is a useful process, the protection of the photosynthetic apparatus by carotenoids is recognized as being essential to the survival of the plant. It is, therefore, interesting at this point to examine the role of carotenoids in the evolution of photosynthetic organisms. It has been proposed that carotenoids originally evolved in order to act as accessory light-harvesting pigments when atmospheric oxygen levels were low, but had the additional benefit of acting as effective antioxidants as these oxygen levels rose. The similar, yet precise, structural arrangements required for efficient light harvesting and photoprotection may have led to pressures for a co-evolutionary process fulfilling both functions.[34]

Much of the work carried out on the photoprotective role of carotenoids in photosynthetic systems has concentrated on the processes in the phototrophic bacteria[32,35,36] and in *in vitro* systems.[33] The conclusions drawn from these studies (see Figure 7) are very relevant to the processes in higher plants and algae where, as yet, very little structural information is available to permit an accurate picture of the mechanisms of photoprotection to be formulated.

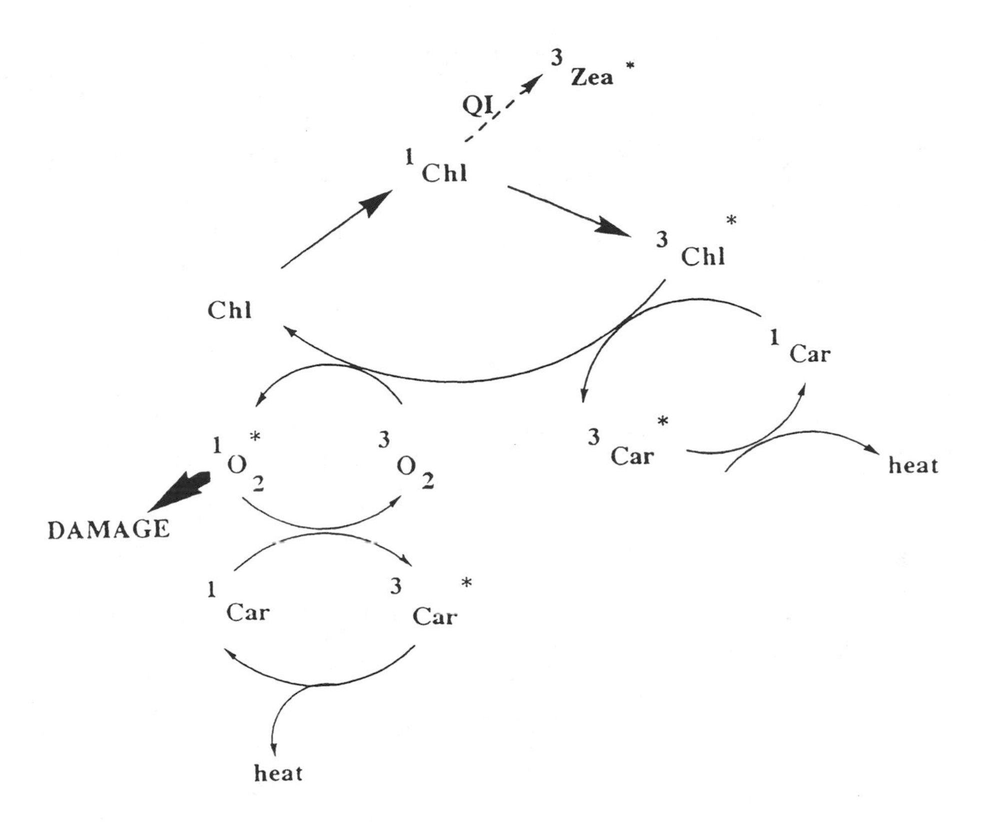

FIGURE 7. Schematic representation of the role of carotenoids in quenching triplet states and in scavenging for singlet oxygen. The possible quenching interaction (QI) of zeaxanthin ($^3Zea^*$) with chlorophyll is also shown.

D. SCAVENGING OF SINGLET-OXYGEN AND RADICAL-OXYGEN SPECIES

Carotenoid can interact with singlet oxygen and free radicals preventing the initiation of potentially lethal processes, such as lipid peroxidation. The mechanism whereby carotenoids such as β-carotene can interact with these reactive compounds has been the subject of much investigation (see Truscott[33] and Krinsky[39]), although there has been very little work carried out in their scavenging role in photosynthetic systems directly.

The majority of the published work on the antioxidant role of carotenoids concerns itself with the properties of β-carotene, although recently reports on the *in vitro* effectiveness of other hydrocarbons (e.g., lycopene[40]) and some of the xanthophylls (e.g., astaxanthin and canthaxanthin[41]) have also been reported.

The energy transfer process from singlet oxygen to β-carotene to produce triplet state β-carotene (Reaction 7) is known to be a very efficient process.[42] To be an effective photoprotector, a carotenoid molecule must meet specific triplet-state energy requirements, i.e., have their lowest-lying excited triplet-state energy less than 94 kJ mol^{-1} (7855 cm^{-1}).[35] Thus, the quenching of singlet oxygen by carotenoids can proceed efficiently only if the triplet energy level of the carotenoid molecule is below that of the $1\Delta g$ state of singlet oxygen. The triplet levels for carotenoids are not well established, with some reports from resonance Raman spectroscopy suggesting a level of 75 kJ mol^{-1}, others at 92 kJ mol^{-1} (isoenergetic with singlet oxygen[42]), or even somewhat lower than 90 kJ mol^{-1}[33]. Carotenoids possessing polyene chains containing nine or more conjugated double bonds were determined, from their triplet energy levels, to be effective scavengers for singlet oxygen.[42] Indeed, carotenoids with seven or fewer conjugated double bonds would not be effective.

This has been observed *in vitro* in solvents,[44] and has also been successfully illustrated in the microorganism *Micrococcus luteus*. Mutant strains of this microorganism, containing carotenoids with fewer than nine conjugated double bonds, were not effective in protecting the organism from photooxidation, and the viability of these cultures was soon lost. The wild-type organism, containing carotenoids with nine conjugated double bonds, remained viable under similar oxidative conditions.[45]

The relationship between the length of chromophore chain (number of conjugated double bonds) and the scavenging ability of the carotenoid molecule is well known.[39] Additional modifications to the simple carotenoid hydrocarbon structure (e.g., introduction of functional groups as in the xanthophylls) can alter the scavenging efficiency of these pigments *in vitro*.[40] This has important implications for the carotenoids found in higher plant and algal chloroplasts, as the bulk of carotenoids are xanthophylls.

$$\text{Chl} + h\nu \longrightarrow {}^{1}\text{Chl}^* \quad (4)$$

$${}^{1}\text{Chl}^* \longrightarrow {}^{3}\text{Chl}^* \quad (5)$$

$${}^{3}\text{Chl}^* + O_2 \longrightarrow \text{Chl} + {}^{1}O_2 \quad (6)$$

$${}^{1}O_2 + \text{Car} \longrightarrow {}^{3}\text{Car}^* + O_2 \quad (7)$$

$${}^{3}\text{Car}^* \longrightarrow \text{Car} + \text{heat} \quad (8)$$

The general scheme for chlorophyll-sensitized production of singlet oxygen and subsequent quenching by carotenoids is shown in Reactions 4 to 8, above. The interaction between the singlet oxygen and β-carotene (Reaction 7) has been found to take place through an exchange electron transfer mechanism,[41] and a similar reaction is also thought to occur with a number of other carotenoids, including lycopene.[40,46,47]

The structure of the β-carotene molecule and of many other carotenoids in possessing an extended series of conjugated double bonds makes them prooxidant in nature and, hence, susceptible to attack by the addition of peroxy radicals.[48,49]

In the presence of oxygen, a new chain-carrying peroxy radical species of β-carotene is formed (Reactions 9 and 10). β-Carotene was found to readily undergo autooxidation, which was found to be highly dependent on the partial pressure of oxygen. Thus, β-carotene would be much more effective at preventing the oxidation of cellular components at low partial pressures of oxygen, and may indeed complement the role of vitamin E (see Chapter 5 by Hess), which is effective at high-oxygen pressures.[49] This protection would, therefore, be of greatest benefit in the conditions found in many mammalian cells where low oxygen concentrations prevail, but the effectiveness of this chain-breaking property in photosynthetic tissues, which generally have high concentrations of oxygen, is not known.

$$\beta\text{-carotene} + ROO\cdot \longrightarrow \beta\text{-car}\cdot \quad (9)$$

$$\beta\text{-car}\cdot + O_2 \longleftrightarrow \beta\text{-car-OO}\cdot \quad (10)$$

It has also been proposed that the β-carotene radical (β-car·) can be removed through interaction with another peroxyl radical (Reaction 11).[48,49] The overall effect of these reactions will be to prevent a series of reactions involving lipid peroxidation by promoting what would be much less serious reactions with the more expendable carotenoid molecules.[49] Importantly, from the point of view of many natural systems where a complex mixture of carotenoids may be present, these unusual antioxidant properties of β-carotene are also probably seen in other carotenoid molecules which possess a series of conjugated double bonds.

$$\beta\text{-Car}\cdot + ROO\cdot \longrightarrow \text{inactive products} \quad (11)$$

It is clear that, *in vitro*, carotenoids have the ability to scavenge singlet oxygen and other highly reactive and potentially lethal oxygen species. The main photoprotective role of these pigments in photosynthetic systems may be through the direct quenching of the triplet-state sensitizer (chlorophyll or bacteriochlorophyll).[50] It was also concluded that, *in vivo*, if carotenoid function were restricted to the scavenging of singlet oxygen, then a significant proportion of bacteriochlorophyll molecules would be irreversibly photooxidized through triplet sensitization. Furthermore, only partial photoprotection would be achieved because (1) the efficiency of this process would depend on the proximity of molecular oxygen to carotenoid and (2) the carotenoid would be competing for the singlet-oxygen energy with other components of the photosynthetic apparatus.[35] Direct quenching of the triplet-state chlorophyll molecule, thus preventing the generation of singlet-oxygen in the first place, provides a much more efficient and effective protective strategy. Certainly the ability of membrane-bound carotenoids to scavenge for exogenous sources of singlet oxygen and free radicals, such as those produced in the presence of atmospheric pollutants, must be severely limited.

E. QUENCHING OF TRIPLET-STATE CHLOROPHYLL

It is now clear that the quenching of triplet-state chlorophyll is probably the single most important process involving carotenoids in photosynthetic systems.

Triplet energy is transferred from the T_1 chlorophyll ($^3Chl^*$) to the ground state carotenoid molecule (^1Car). Subsequently, nonradiative energy dissipation occurs by intersystem crossing of the T1 carotenoid ($^3Car^*$) produced in this reaction. This process fits in well with the information available concerning the relative energy levels of these components, although the precise mechanisms of the above reactions are now known. Reaction 1_2 probably occurs through an electron exchange mechanism.[32] Other substrates, especially oxygen, will compete for this excitation energy, and it has been calculated that photoprotection through triplet-triplet energy transfer will only be effective when the rate constant for this reaction is at least ten times greater than that for the reaction between the T_1 chlorophyll molecule and molecular oxygen.[51]

$$^3Chl^* + {}^1Car \longrightarrow {}^1Chl + {}^3Car^* \tag{12}$$

$$^3Car^* \longrightarrow {}^1Car + heat \tag{13}$$

The main function of the reaction-center carotenoid, therefore, is to accept the triplet energy of the excited chlorophylls and dissipate it harmlessly as heat. In photosynthetic membranes, this process is highly dependent on the proximity of the pigment molecules within the pigment-protein complexes.[36,50] A precise molecular arrangement of pigment molecules within the pigment protein complexes is required for effective photoprotection. *In vitro,* where detergents increase the intermolecular distance of pigments in the main light-harvesting complex, the rate and extent of photobleaching of chlorophyll are increased.[52]

In the phototrophic bacteria, a specific role for the 15-*cis*-carotenoids in the reaction center has been proposed.[32,53] This 15-*cis* configuration is thought to be more suitable for energy dissipation in the reaction center than the other possible *cis-trans* configurations, probably due to the properties of the triplet state. In contrast, in higher plants, the data obtained strongly suggests that the β-carotene molecules associated with the D1-D2 polypeptide are in the all-*trans* configuration. The energy dissipation process in plants may involve *trans-cis* isomerization of β-carotene at the triplet level.[54]

Detailed studies using pico-second time-resolved resonance Raman spectroscopy have elucidated many of the processes involved in photoprotection by carotenoids in reaction centers of the phototrophic bacteria (see Reference 32). This system has the advantage of being much simpler, in terms of both the pigments and of the polypeptide structure, than that found in higher plants and algae. However, many of the details concerning the precise mechanism of photoprotection are still unknown.[32]

F. THE XANTHOPHYLL CYCLE

The harmless dissipation of excess excitation energy absorbed by the photosynthetic apparatus through processes involving the carotenoid zeaxanthin has a major photoprotective effect in higher plants and in some algae. This topic is reviewed at length in Chapter 4.

V. EFFECT OF PHOTOBLEACHING

A. PHOTOINHIBITION

Under optimum conditions, the carotenoids associated with the photosynthetic apparatus will serve to protect the chloroplast from photooxidative damage through the processes described above. In many cases, these protective systems become overloaded, resulting in damage to the components of the thylakoid membrane.

The initial exposure to high light intensity may induce some subtle changes in the carotenoid content and composition of the photosynthetic tissues. These may involve the photoisomerization of certain carotenoids[55] and the induction of the xanthophyll cycle (see Chapter 4). The occurrence of geometrical isomers of carotenoids in photosynthetic tissues has been reported for some time and is of great interest, as their detection may lead to a greater understanding of the precise mechanisms concerning light harvesting and photoprotection.

Illumination of leaf tissues, chloroplasts, or thylakoids with high light intensities under aerobic conditions may result in the rapid bleaching of photosynthetic pigments. Such bleaching reflects oxidative damage in these systems.[56] The photobleaching of chlorophyll in the light-harvesting complexes can be considered to be a self-sensitized type II reaction.[52] These bleaching processes are highly dependent on the presence of oxygen, and can be controlled through the action of various scavengers of singlet oxygen and promoted in the presence of D_2O (which extends the life time of singlet oxygen). Under anaerobic conditions, bleaching of pigments is minimal, i.e., it is the combination of light and oxygen that is most damaging.

The photobleaching of pigments generally occurs in a well-defined order. The carotenoids are destroyed at a faster rate than the chlorophylls, with the reaction center-bound β-carotene being particularly susceptible to photooxidative conditions. Neoxanthin is also very susceptible to these conditions, and is lost much more readily than the other xanthophylls, including violaxanthin and lutein. The rate of loss of all of the xanthophyll cycle components (violaxanthin, anteraxanthin, and zeaxanthin) is similar *in vivo*. Whereas zeaxanthin is readily formed in high light leaves, in *in vitro* systems (isolated chloroplasts and thylakoids) exposure to light will not normally induce the xanthophyll cycle unless additional ascorbate is present.

Lutein is the major carotenoid component in the chloropast, and is, under most conditions, particularly stable to photooxidative degradation. Chlorophyll a is generally destroyed at a faster rate than chlorophyll b. Taking into account the changes in the xanthophyll cycle pigments, identical patterns of bleaching are seen *in vivo* and *in vitro*. This bleaching seen in isolated chloroplasts and thylakoids as a consequence of photoinhibitory conditions can be stimulated in the presence of D_2O (which can extend the life time of singlet oxygen up to tenfold), and can be prevented in the presence of ascorbate which acts as an effective scavenger of oxidizing species. An almost identical pattern of pigment loss is seen here as in illuminated control tissues, suggesting that a similar oxidative process, probably involving singlet oxygen is involved.[73]

In isolated chloroplasts, photodestruction of pigments in all complexes was found to occur as a result of exposure to high light intensities.[57] The relative pigment content of LHC II complex was reduced to a greater extent than any other complex. Pigment levels in the CC I complex were also greatly reduced. This resulted in an increase in the levels of chlorophylls and carotenoids in the free-pigment miscells following illumination. Similar data were obtained in chloroplasts obtained from "greened" etiolated seedlings. In "greened" seedlings, the extent and rate of pigment loss were greater than that seen in light-grown plants. An almost identical pattern of pigment photobleaching has been seen in the presence of dichlorpophenyl-dimethylurea (DCMU) for chloroplasts and with monuron in intact leaves[57] (Figure 8).

In isolated LHC I that was exposed to 200 Klux, both chlorophylls and β-carotene were rapidly degraded, with the xanthophylls remaining much more stable.[58] A greater loss of β-carotene than either of the chlorophylls, indicated a major protective role for this carotenoid in this antenna complex. The xanthophyll pigments appeared to provide only a structural role in maintaining the integrity of the complexes rather than a photoprotective one. Isolated CC I (also exposed to 200 Klux) showed very rapid losses of chlorophyll b, compared to β-carotene, chlorophyll a, or the remaining xanthophylls. Indeed, the high stability of chlorophyll a in this complex was probably due to its very high level of β-carotene. The low xanthophyll levels in CC I afforded little protection, if any, to chlorophyll b.

In a study on isolated thylakoids and in PS I and PS II preparations carotenoids were found to be bleached to a greater extent in PS I preparations.[59] This was initially characterized by a slow rate of loss, indicating the presence of energy traps which serve to protect the bulk pigments. The core-antenna complexes were also more sensitive to illumination than the peripheral complexes.

1. Formation of β-Carotene 5,6-Epoxide

In a series of *in vitro* experiments, Lee and Britton[95] showed that the formation of β-carotene-5,6-epoxide (5,6-epoxy-5,6-dihydro-β,β-carotene, Figure 9) was a reliable indicator of oxidative damage to β-carotene. Although it has been reported to be widely present in flowers and fruit, this epoxide has not yet been detected as a natural component of healthy, undamaged photosynthetic tissues, and has only been found as a consequence of oxidative damage to leaves or isolated chloroplasts. A correlation between the observed photobleaching of β-carotene in thylakoid membranes and the occurrence of all-*trans*-β-carotene monoepoxide has been suggested.[60] Furthermore, the appearance of this epoxide in spinach thylakoids and in the cyanobacterium *Synechococcus vulcanus* was proposed to be largely the result of an enzymic epoxidation/de-epoxidation cycle comparable with the xanthophyll cycle. The production of β-carotene-5,6-epoxide was thought to be based on a diurnal cycle, i.e., levels of the epoxide were low following a dark period and rose following exposure to natural daylight. This sequence is, in fact, the reversal of xanthophyll cycle in which it is the de-epoxidation reactions which are promoted in the light.

In a more recent study,[61] neither β-carotene-5,6-epoxide nor its furanoid derivative β-carotene-5,8-epoxide (mutatochrome) were detected in the leaves of a number of higher plant species or algae, except under conditions of oxidative stress. These studies also confirmed by spectroscopic analysis that the epoxide produced

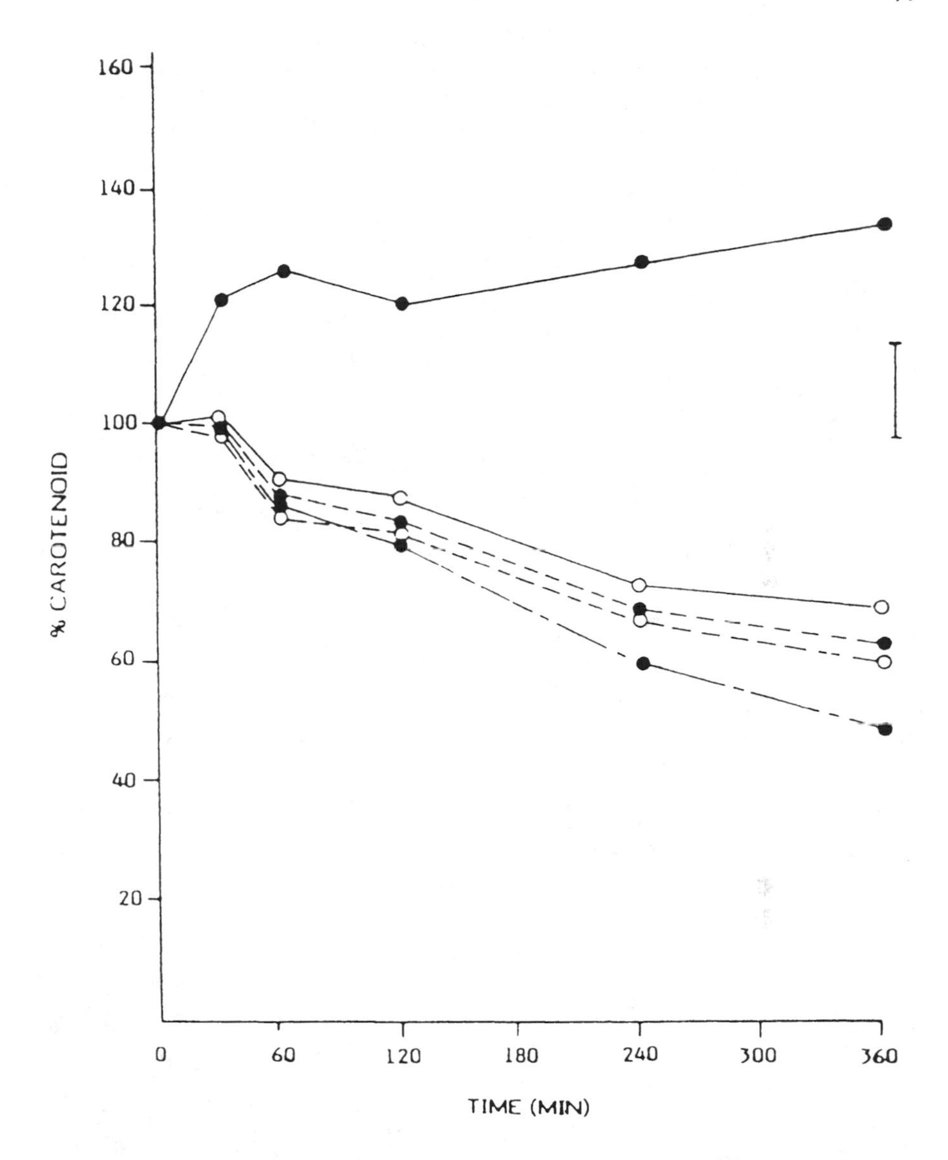

FIGURE 8. Photodestruction of carotenoids in isolated chloroplasts from *Hordeum vulgare*. The values shown are a mean of at least three separate determinations: β-carotene-5,6-epoxide (●—●), lutein (○—○), violaxanthin (●--●), neoxanthin (○--○), β-carotene (●—●). (From Barry, P., Young, A. J., and Britton, G., *J. Exp. Bot.*, 41, 123, 1990. With permission.)

in these reactions was optically inactive, and, therefore, not produced enzymically, but must have been derived from β-carotene by purely photochemical or chemical reactions to give a mixture of the (5*R*, 6*S*) and (5*S*, 6*R*) isomers.

Conflicting data exists concerning the location of the β-carotene-5,6-epoxide produced in chloroplasts or thylakoids. Preferential location in PS II particles has

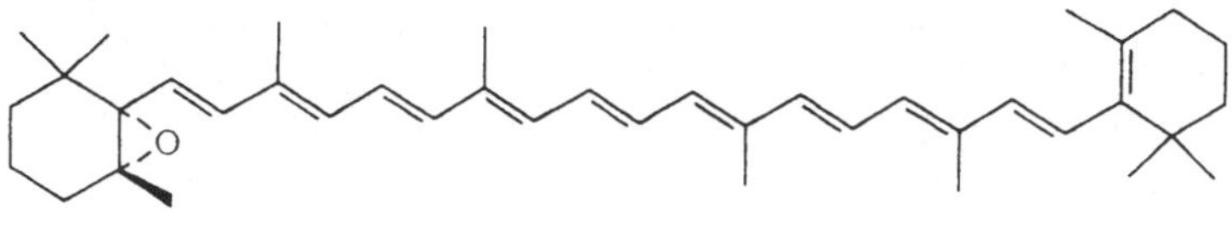

FIGURE 9. β-Carotene-5,6-epoxide.

been reported,[60] and high levels in the PS I-reaction center complexes has been observed.[61] This latter observation could only detect the epoxide in the PS II complexes following very harsh treatment of thylakoids with DCMU.

B. EFFECTS OF HERBICIDES

The action of certain herbicides on photosynthetic tissues provides the best examples of carotenoids protecting plants from photooxidation. Herbicides can act upon the chloroplast pigments of plants in two ways: (1) they can either destroy the plants existing pigments, generally through the production of oxidative species such as singlet oxygen and superoxide (e.g., DCMU, paraquat), or (2) they can interfere with the biosynthesis of these pigments (e.g., diflufenican, amitrole). A limited number of herbicidal compounds may act in both ways (e.g., certain of the diphenyl ethers). Damage is brought about both by a lack of energy dissipation when the electron flow through PS II or cyclic PS I becomes limited (e.g., DCMU), and, with certain compounds, by the direct production of toxic oxygen species (e.g., the action of paraquat).

1. Inhibition of Carotenoid Biosynthesis

A number of reviews have recently been published concerning the mode of action of those compounds that inhibit carotenogenesis, and it is not the intention of this chapter to cover much of this information.[62-65] The vital role that carotenoids play in photoprotection is well illustrated in two *in vivo* systems: (1) a range of mutants (microorganisms and higher plants) deficient in their normal carotenoid composition (see Section IV., above), and (2) the effect of inhibiting the biosynthesis of carotenoids in photosynthetic systems.

The inhibition of carotenoid biosynthesis has two main effects: first, the synthesis of the normal cyclic carotenoids (e.g., β-carotene, lutein, etc.) is prevented through the inhibition of the desaturase and cyclization reactions in the biosynthetic pathway, resulting in the accumulation of a number of precursors (e.g., phytoene) (see Section II., above). Second, the absence of these cyclic carotenoids prevents the normal assembly of the photosynthetic apparatus so that the resulting plastids formed following treatment with these inhibitors are generally either inactive or show poor photosynthetic performance. Recent reconstitution studies have shown an absolute requirement for xanthophylls in the successful reassembly of isolated LHC II.[66]

A number of studies have now shown that plants treated with inhibitors of the desaturation and cyclization reactions of carotenoid biosynthesis are readily susceptible to photooxidation. These inhibitors, unlike those that interfere with electron transport processes, do not have a requirement for light for their action, and the plants produced following herbicide treatment are usually extensively bleached white, or in the case of inhibitors of cyclization (e.g., amitrole), yellow.

Plants treated with norflurazon and other bleaching compounds can form fully developed and functional chloroplasts (classed as rudimentary sun-type chloroplasts) in low light, whereas in higher light intensities, the plants are bleached white and contain no carotenoid or chlorophyll pigments.[67] Similarly, transfer of treated plants grown in low light conditions, which contain chlorophyll and have green leaves, to higher light intensities will result in the rapid bleaching of the accumulated chlorophylls, as a result of the absence of the protective cyclic carotenoids. Similar observations have been reported in wheat seedlings treated with fluridone or norflurazon.[68] The absence of the cyclic carotenoids results in a lack of effective energy dissipation processes through direct means via the carotenoids, and indirectly, through interference with the structural arrangement of the pigment-pigment and pigment-protein associations.

Rye seedlings grown at very low light intensities in the presence of a number of inhibitors were pale green in color, and contained reduced amounts of carotenoids and chlorophylls compared to the control plants.[69] Those seedlings treated with metflurazon (SAN 6706) or difunon showed substantial bleaching of these pigments upon transfer of the plants to a higher light intensity. This photodestruction of pigment required oxygen, and was accompanied by the degradation of a range of cellular components, including the 70S ribosomes and chloroplast enzymes.[70] In those seedlings treated with amitrole or haloxydine, the chlorophylls appeared to be quite stable. This effect could not be seen *in vitro,* with chloroplasts isolated from low light grown plants treated with metflurazon.[71] Other herbicides which allowed chlorophyll formation at low light but bleaching at high light intensities include pyrichlor and dichlormate.[72] Examination by transmission electron microscopy of plants treated with inhibitors of carotenoid biosynthesis showed that high light grown plants had extensive destruction of chloroplast components.[64]

Although, in theory, the photochemical properties of some accumulated intermediates should enable them to be able to act as effective scavengers of singlet oxygen (e.g., lycopene with 13 conjugated double bonds), it clearly does not happen *in vivo.* It is believed that these intermediates are not located in suitable sites for effective control of oxidative species, since a close proximity between the carotenoid molecule and singlet oxygen is essential for efficient scavenging. A similar relationship is needed between carotenoid and chlorophyll molecule for effective quenching of the chlorophyll triplet state.

2. Interference with Electron Transport Processes

Many herbicidal compounds will affect the photosynthetic pigments of treated tissues through the generation of highly reactive oxygen species (e.g., superoxide generation by paraquat) which result in extensive chlorosis and necrosis of leaf tissues.[62] A study of the effect of several compounds representing a number of different classes of herbicides on chloroplast pigments *in vivo* and *in vitro* has recently been undertaken.[57,73] Inhibitors of electron transport at PS II have an effect similar to the effects of exposure to high light, both in isolated chloroplasts (DCMU) and in intact leaves (monuron). *In vivo,* DCMU and monuron both inhibit the de-epoxidation of violaxanthin to zeaxanthin.

The application of other compounds with different modes of action, e.g., the diphenyl ethers (e.g., acifluorfen, fomasafen) and paraquat to intact leaves shows slightly different patterns of pigment composition compared to controls (Figures 10A and 10B). Either different oxidizing species are involved (e.g., superoxide with paraquat treatment) and/or the oxidizing species are produced at different sites in the plant cell or chloroplast. The membrane-bound carotenoids will only be effective in scavenging for oxidizing species that are produced in close proximity.

The application of paraquat to intact leaves in the light causes very rapid and severe chlorosis of tissues. The pattern of pigment loss is similar to that seen with DCMU, except that the presence of paraquat appears to induce violaxanthin de-epoxidase at low light intensities, leading to the accumulation of zeaxanthin.[73] The diphenyl-ether compounds also cause rapid bleaching, although close examination shows that the kinetics of carotenoid and chlorophyll loss are quite different.[73] Unlike any of the bleaching processes described above, acifluorfen and fomasafen affected chlorophyll degradation at a much faster rate than they did degradation of carotenoids. Of the carotenoids, the xanthophylls are much more readily destroyed than β-carotene, suggesting that the reaction-center bound β-carotene is not as greatly involved in photoprotection in these tissues as it is in photoinhibitory conditions or with other herbicides. Violaxanthin and neoxanthin are particularly susceptible to the oxidative conditions in the presence of the diphenyl-ethers. One of the most noticeable features of the action of these herbicides is the production of large amounts of xanthophyll acyl esters (see below) following prolonged treatment. Paraquat and certain diphenyl ethers serve to induce the xanthophyll cycle so that small amounts of zeaxanthin may be found in treated tissues, although it is not known whether they directly promote the de-epoxidation of violaxanthin. It is more likely that they have a secondary effect that will affect the epoxidation state of the leaves (e.g., alteration of the *trans*-thylakoid pH gradient.[73] The exact nature of the oxidative species produced by the diphenyl ethers is not known, although increased production of singlet oxygen in the thylakoids is likely. There is no clear evidence that either superoxide or H_2O_2 are formed as a result of the application of diphenyl ethers to plant tissues. The diphenyl ether herbicides were originally thought to interact directly with carotenoids in the thylakoid membrane, and so become sensitized, resulting in the photooxidation of cellular components. It is now known that this is not the case, and porphyrin compounds (e.g., protoporphyrin IX) are the sensitizers produced by the inhibition of protoporphyrinogen oxidase by diphenyl ethers.

C. OTHER PHOTOOXIDATIVE PROCESSES

Two distinct patterns of pigment bleaching are found in plant tissues. The majority of stress treatments (exposure to high light intensities, herbicide treatment, etc., see above) will result in greater damage to the carotenoids than to the chlorophylls. In these cases, the carotenoids are probably in the first line of defense as quenchers of triplet state chlorophyll and of singlet oxygen produced in the thylakoid membrane. Another commonly observed pattern of pigment loss may be observed in which the chlorophylls are degraded at a faster rate than the carotenoids. This may be partly due to enzymic degradation, although purely oxidative conditions may produce similar results.

1. Atmospheric Pollutants

The effects of photooxidative processes on plant tissues should not be confused with the effects of nutrient deprivation, and, in some cases, toxicity on the assembly of the thylakoid membrane and its components. For example, iron deficiency in pea will result in the formation of almost yellow-colored leaves with a very low chlorophyll content.[74] Similar observations have been made for the needles of a number of tree species,[75,76] and it may be difficult, therefore, to differentiate between these conditions and photooxidative destruction simply by examining the pigment compositions of leaves and needles.

The loss of pigments in leaves and needles is one of the most noticeable features in trees subjected to atmospheric pollution. The series of events involved in such damage has not, as yet, been fully elucidated, although it is clear that photooxidative destruction of chloroplast components occurs in many cases. The effects of a number of pollutants (ozone, NO_2, and SO_2) on the carotenoid composition of leaves and needles have been investigated. In the case of ozone-mediated damage to leaves and needles, in which the production of hydroxyl radicals is thought to be important,[77] damage to photosynthetic tissues is readily observed. A detailed examination of the changes in pigment composition in young barley seedlings as a result of exposure to ozone and NO_2 has recently been undertaken.[78] The visible symptoms of fumigation with these pollutants are quite different: ozone resulted in the appearance of chlorotic patches all over the leaves, whereas NO_2 caused necrotic lesions to appear specifically at the edge of the leaf. Differences were also seen in the pattern of pigment bleaching with these two pollutants. The destruction of chlorophyll a and b proceeded at similar rates in the presence of ozone, but chlorophyll b levels, in particular, were severely reduced with NO_2 treatment. The pattern of carotenoid destruction was similar with both pollutants, with neoxanthin being the most susceptible and β-carotene the most stable. The main xanthophyll, lutein, was much less sensitive to NO_2 treatment than to ozone. Traces of xanthophyll acyl esters were detected following prolonged treatment with these pollutants.

Changes in the pigment content and composition of leaves and needles to predict damage to photosynthetic systems has been proposed.[79] Analysis of the pigment content of leaves taken from trees in field sites where ozone was thought to be the major pollutant suggested that long-term exposure to low concentrations of ozone may serve to promote the epoxidation of zeaxanthin to antheraxanthin and zeaxanthin. An increase in the violaxanthin content of needles from young *Picea abies* trees when exposed to a combination of ozone and acid mist has been observed.[80] Such changes could form part of a regulatory process leading to protection from light-mediated stress. Considerable variation in the violaxanthin and zeaxanthin contents of damaged needles of *P. abies,* particularly with needle age has been

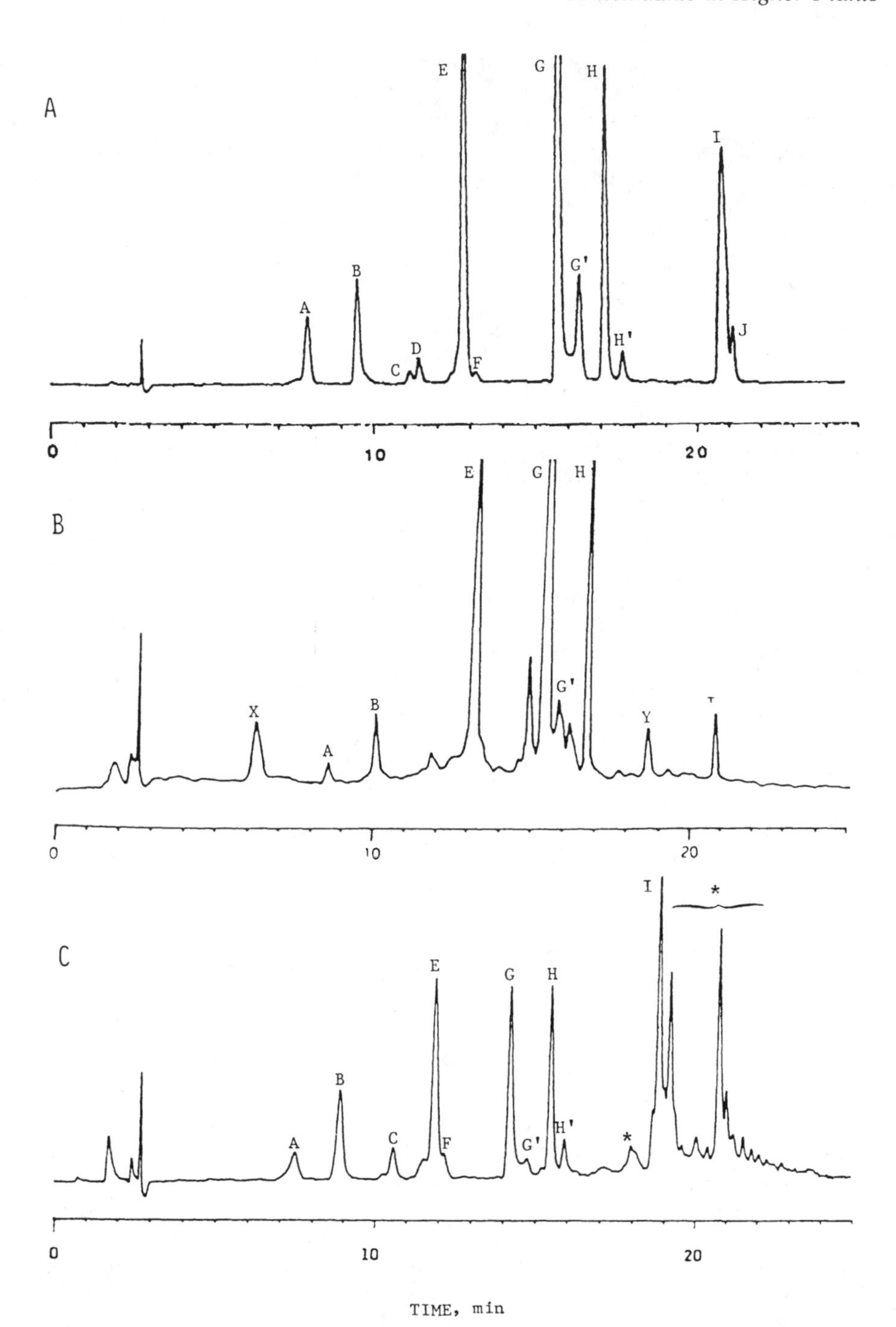
A
E
G
H
I
B
G'
A
D
H'
J
C
F
0
10
20
B
E
G
H
G'
X
B
Y
A
0
10
20
I
*
C
E
G
H
B
H'
A
C
F
G'
*
0
10
20
TIME, min

FIGURE 10.

FIGURE 10 (on page 82). Chromatograms of pigment extracts from leaves of *Hordeum vulgare*. A reversed-phase HPLC system yielded separations of the major pigments that were detected at 447 nm.[57] The relative detector response is plotted as a function of elution time for three samples: (A) Control leaf sample — pigment identification: (A) neoxanthin; (B) violaxanthin; (C) lutein-5,6-epoxide; (D) antheraxanthin; (E) lutein; (F) zeaxanthin; (G) chlorophyll b; (G′) chlorophyll b isomer; (H) chlorophyll a; (H′) chlorophyll a isomer; (I) all *trans*-β-carotene; and (J) *cis*-β-carotene. (B) Leaf sample from paraquat-treated plants. Tissue was extensively bleached, and levels of all pigments were reduced compared to controls. Peak X, a chlorophyll degradation product, and peak Y, β-carotene-5,6-epoxide, were identified in these extracts. (C) Leaf sample from rose-bengal treated plants. This pigment composition was similar to that in naturally senescing leaves and in tissue from plants treated with the herbicides, acifluorfen and fomasafen.[85] The asterisk (*) denotes a complex mixture of xanthophyll acyl esters, of which the main component has been shown by MS to be lutein-*bis*-linolenate.[86]

recorded.[81] Zeaxanthin levels were highest in the partially necrotic needles. In a more recent study, a shift in the epoxidation state of these xanthophylls in ozone-treated barley seedlings could not be observed.[78] The possibility of using the presence of β-carotene-5,6-epoxide in leaves and needles as an indicator of specific photooxidative damage to the thylakoid has been proposed.[61] Young *P. abies* and *Pinus sylvestris* trees that were subjected to elevated levels of ozone and sulfur dioxide lacked visible signs of photobleaching. These trees had no alteration in their carotenoid or chlorophyll composition,[80] suggesting that the pigment content and composition could not be used as a predictor of oxidative stress in these species.

2. Exogenous Sources of Singlet Oxygen

A number of photosensitizer dyes can be used to produce singlet oxygen in the presence of light and oxygen. The action of one of these sensitizers, rose bengal, on the photooxidative breakdown of photosynthetic tissues has been investigated.[83-85] These studies showed that, in the presence of rose bengal, bleaching of pigments either *in vitro* or in illuminated leaf discs proceeds very rapidly. The effect of treatment of leaves with rose bengal is shown in Table 2. Such treatment increased the ratios of chlorophyll a to b and of total carotenoid to total chlorophyll. The ratio of lutein to β-carotene was severely reduced, indicating a relatively high stability of β-carotene compared to the xanthophylls, and, indeed, to the chlorophylls. This situation is the reverse of that found in high light treatments. Exposure of leaves to rose bengal in the light also results in the formation of very high amounts of xanthophyll acyl esters (see below). Similar results were obtained with some of the diphenyl ether herbicides (e.g., acifluorfen-methyl and fomasafen).

3. Formation of Xanthophyll Acyl Esters

The occurrence of xanthophyll acyl esters in photosynthetic tissues has only recently been reported. These compounds have not been detected in leaves maintained under optimum conditions. Their presence may be restricted to tissues subjected to oxidative damage, e.g., treatment of leaves with the singlet-oxygen sensitizer, rose bengal, (Figure 10c) and in leaves treated with certain of the diphenyl ether herbicides (acifluorfen, etc.),[55] drought-stressed leaves of barley,[86] and in dark-induced senescent leaves.[87] They are not, however, universally found in photooxidized tissues, and have not been detected as a result of prolonged exposure to high light intensities or to other herbicides (e.g., monuron and paraquat). Their

TABLE 2
Carotenoid Composition of Leaves of *H. vulgare* Subjected to Various Forms of Oxidative Stress

	Control	High-light[b] conditions	Paraquat[b] treatment	Monuron[b] treatment	Rose-bengal[b] treatment
Neoxanthin	12.4	11.9	16.1	12.4	4.6
Violaxanthin	20.1	8.1	10.8	17.6	9.9
Anteraxanthin	1.1	11.0	2.5	0.7	3.1
Lutein	38.0	48.0	67.2	55.0	13.8
Zeaxanthin[a]	—	8.6	—	—	2.0
β-Carotene-5,6-epoxide[a]	—	1.6	+ −	1.7	—
β-Carotene	28.4	20.7	3.4	12.6	23.3
Xanthophyll acyl esters	—	—	—	—	42.9
Ratio chlorophyll a to b	2.30:1	1.94:1	1.20:1	1.42:1	2.89:1
Ratio carotenoid to chlorophyll	0.51:1	0.34:1	0.23:1	0.33:1	0.97:1
Ratio lutein to β-carotene	1.33:1	2.32:1	19.82:1	4.40:1	0.59:1

Note: Data are presented as a percent of total carotenoids in the extract as determined from HPLC.

[a] Zeaxanthin and + β-carotene-5,6-epoxide synthesis stimulated in early stages of paraquat treatment.
[b] High-light (singlet oxygen); paraquat (superoxide); monuron (singlet oxygen and hydroxyl radicals); and rose-bengal (exogenous singlet oxygen).

From Barry, P., Young, A. J., and Britton, G., *J. Exp. Bot.*, 41, 123, 1990. With permission.

formation may be promoted by certain oxidative conditions. It has been suggested that two factors are important in the production of these xanthophyll acyl esters in photosynthetic tissues: namely the nature of the oxidizing species involved and their site of production.[55] It is tentatively proposed that an extra-chloroplastic source of oxidizing agents, involving extensive lipid peroxidation, may be responsible.

The production of these esters was found to be identical under light (drought) and dark (senescence) conditions, suggesting that similar degradative processes are involved. Unsaturated fatty acids in the thylakoid membrane are particularly prone to attack by free radical species. The resulting lipid peroxidation leads to disruption of the membrane, releasing carotenoid and lipid into the cytosol. Free fatty acids liberated from lipid degradation will then be available to esterify the xanthophylls that have been released from the degraded pigment-protein complexes. The location of these esters is thought to be within plastoglobuli that also contain free, unesterified, carotenoids (e.g., β-carotene and lutein),[87,88] and possibly chlorophyll.[89] An important observation was that esters of all major chloroplast xanthophylls were detected in stressed tissues, the main component being lutein-*bis*-linolenate.[86] Data indicated that these xanthophyll esters were produced in approximately the same proportions as that of the carotenoids content in the thylakoid membrane.

VI. ADAPTATION TO ENVIRONMENTAL STRESS

It has become clear that one of the main mechanisms whereby plants can adapt to changing environmental conditions is through the xanthophyll cycle. Many plants show quite large differences in the absolute levels of these carotenoids and in their ability to synthesize zeaxanthin in direct response to stress conditions. This major protective response is discussed at length in Chapter 4.

In addition to major changes in the xanthophyll cycle pigments in sun- and shade-adapted leaves (see Chapter 4), the ratio of α-carotene to β-carotene also differed.[90] α-Carotene levels were found to be much higher in shade plants grown in the shade or in the sun. In contrast, little or no α-carotene was found in sun leaves. Similarly, those plants containing high levels of α-carotene were shown to have a much higher ratio of α-carotene to β-carotene when grown in low light conditions.[90] Needles of *P. abies* from the shade had a much higher proportion of total carotene as α-carotene than did sun-grown needles.[91] A constant ratio of α-carotene to β-carotene in the pigment-protein complexes of higher plants has been found, suggesting a similar functional role for both these carotenoids.[92] α-Carotene is an intermediate in the synthesis of the main light-harvesting carotenoid, lutein, whereas β-carotene is the precursor of neoxanthin, violaxanthin, antheraxanthin, and zeaxanthin. The relevant contents of α-carotene and β-carotene in sun and shaded leaves may have no functional significance, and may simply reflect the contents of lutein and the remaining xanthophylls in these plants, the levels of which also vary considerably in these plants.[93]

Changes in the activities of active oxygen scavengers in spinach in response to cold stress under conditions of excess light have recently been examined.[94] Only minor alterations in the levels of neoxanthin and of β-carotene were recorded, but the levels of the xanthophyll-cycle carotenoids were found to be much higher in the hardened plants compared to the control plants.

VII. CONCLUSIONS

Carotenoids play an important role in all photosynthetic organisms in protecting them from photooxidative damage. Although these pigments have an efficient mechanism for scavenging for the damaging oxygen species directly, especially singlet oxygen, a much more efficient protective process involves preventing the generation of these species in the first place. It is also important to consider that carotenoids have an important role in providing the structural integrity of the photosynthetic apparatus.

REFERENCES

1. **Qureshi, N. and Porter, J. W.,** Conversion of acetyl-*Co*-enzyme A to isopentenyl pyrophosphate, in *Biosynthesis of Isoprenoid Compounds*, Porter, J. W. and Spurgeon, S. L., Eds., John Wiley & Sons, New York, 1981, 47.
2. **Poulter, D. and Rilling, H. C.,** Prenyl transferases and isomerase, in *Biosynthesis of Isoprenoid Compounds*, Porter, J. W. and Spurgeon, S. L., Eds., John Wiley & Sons, New York, 1981, 161.

3. **Dogbo, O. and Camara, B.,** Purification of isopentenyl pyrophosphate isomerase and geranyl geranyl pyrophosphate synthase from *Capsicum* chloroplasts by affinity chromatography, *Biochim. Biophys. Acta,* 920, 140, 1987.
4. **Lutzow, M. and Beyer, P.,** The isopentenyl-diphosphate Δ-isomerase and its relation to the phytoene synthase complex in daffodil chromoplasts,Biochim. Biophys. Acta, 959, 118, 1988.
5. **Britton, G.,** Biosynthesis of carotenoids, in *Plant Pigments,* Goodwin, T. W., Ed., Academic Press, London, 1988, chap. 3.
6. **Dogbo, O., Bardat, F., Laferriere, A., Quennemet, J., Brangeon, J., and Camara, B.,** Metabolism of plastid terpenoids. I. Biosynthesis of phytoene in plastid stroma from higher plants, *Plant Sci.,* 49, 89, 1987.
7. **Britton, G.,** Carotenoid biosynthesis — an overview, in *Carotenoids, Chemistry and Biology,* Krinsky, N. I., Mathews-Roth, M., and Taylor, R. F., Eds., Plenum Press, New York, 1990, 167.
8. **Bramley, P. M.,** Carotenoid biosynthesis, in *Methods in Plant Biochemistry, Vol. 8,* Lea, P. J., Ed., Academic Press, London, in press.
9. **Dogbo, O., Laferriere, A., Harlingue, A., and Camara, B.,** Carotenoid biosynthesis: isolation and characterisation of a bifunctional enzyme catalyzing the synthesis of phytoene, *Proc. Natl. Acad. Sci. U.S.A.,* 85, 7054, 1988.
10. **Britton, G.,** Carotenoid biosynthesis — a target for herbicide activity, *Z. Naturforsch.,* 34c, 979, 1979.
11. **Lutke-Brinkhaus, F., Liedvogel, B., Kreuz, K., and Kleinig, H.,** Phytoene synthase and phytoene dehydrogenase associated with envelope membranes from spinach chloroplasts, *Planta,* 156, 176, 1982.
12. **Grumbach, K. H. and Britton, G.,** Carotenoid localisation and biosynthesis in radish seedlings (*Raphanus sativus*) grown in the presence or absence of bleaching herbicides, in *Advances in Photosynthesis Research,* vol. 4, Sybesma, C., Ed., Martinus Nijhoff/Dr. W. Junk Publishers, The Hague, 1984, 69.
13. **Kreuz, K., Beyer, P., and Kleinig, H.,** The site of carotenogenic enzymes in chromoplasts from *Narcissus pseudonarcissus, Planta,* 154, 66, 1982.
14. **Camara, B., Bardat, F., and Moneger, R.,** Sites of biogenesis of carotenoids in *Capsicum chromoplasts, Eur. J. Biochem.,* 127, 255, 1982.
15. **Britton, G.,** The biosynthesis of carotenoids: a progress report, *Pure Appl. Chem.,* 63(1), 101, 1991.
16. **Beyer, P., Mayer, M., and Kleinig, H.,** Molecular oxygen and the state of geometric isomerism of intermediates are essential in the carotene desaturation and cyclization reactions in daffodil chromoplasts, *Eur. J. Biochem.,* 184, 141, 1989.
17. **Beyer, P. and Kleinig, H.,** On the desaturation and cyclization of carotenes in chromoplast membranes, in *Carotenoids, Chemistry and Biology,* Krinsky, N. I., Mathews-Roth, M., and Taylor, R. F., Eds., Plenum Press, New York, 1990, 195.
18. **Camara, B. and Dogbo, O.,** Demonstration and solubilization of lycopene cyclase from *Capsicum* chromoplast membranes, *Plant Physiol.,* 80, 172, 1986.
19. **Camara, B. and Moneger, R.,** Biosynthetic capabilities and localization of enzymatic activities in carotenoids metabolism of *Capsicum annum* isolated chromoplasts, *Physiol. Veg.,* 20(4), 757, 1982.
20. **Demmig-Adams, C.,** Carotenoids and photoprotection in plants: a role for the xanthophyll zeaxanthin, *Biochim. Biophys. Acta,* 1020, 1, 1990.
21. **Siefermann-Harms, D.,** Carotenoids in photosynthesis. I. Location in photosynthetic membranes and light-harvesting function, *Biochim. Biophys. Acta,* 811, 325, 1985.
22. **Codgell, R.,** The function of pigments in chloroplasts, in *Plant Pigments,* Goodwin, T. W., Ed., Academic Press, London, 1988, 183.
23. **Joyard, J., Block, M. A., and Douce, R.,** Molecular aspects of plastid envelope biochemistry, *Eur. J. Biochem.,* 199, 489, 1991.
24. **Grumbach, K. H.,** Distribution of chlorophylls, carotenoids and quinones in chloroplasts of higher plants, *Z. Naturforsch.,* 38c, 996, 1983.

25. **Douce, R. and Joyard, J.,** Isolation and properties of the envelope of spinach chloroplasts, in *Plant Organelles,* Reid, E., Ed., Ellis Horwood, Chichester, 1979, 47.
26. **Barry, P. and Pallett, K. E.,** Herbicidal inhibition of carotenogenesis detected by HPLC, *Z. Naturforsch.*, 45c, 492, 1990.
27. **Chu, Z. X. and Anderson, J. M.,** Modulation of the light-harvesting assemblies in chloroplasts of a shade plant, *Alocasia macrorrhiza, Photobiochem. Photobiophys.*, 8, 1, 1984.
28. **Lichtenthaler, H. K., Buschman, C., Doell, M., Fietz, H.-J., Bach, T., Kozel, U., Meier, D., and Rahmsdorf, U.,** Photosynthetic activity, chloroplast ultrastructure and leaf characteristics of high light and low light plants of sun and shade leaves, *Photosynth. Res.*, 2, 115, 1981.
29. **Barry, P. and Pallett, K. E.,** The effect of diflufenican on pigment levels in carotenogenic systems, *Proc. EWRS Symp. Factors Affecting Herbicidal Activity and Selectivity,* 51, 1988.
30. **Gillbro, T. and Cogdell, R. J.,** Carotenoid fluorescence, *Chem. Phys. Lett.*, 158, 312, 1989.
31. **Bondarev, S. L., Bachilo, S. M., Dvornikov, S. S., and Tikhomirov, S. A.,** $S_2 \rightarrow S_0$ fluorescence and transient $S_n \leftarrow S_1$ absorption of all-trans β-carotene in solid and liquid solutions, *J. Photochem. Photobiol. A,* 46, 315, 1989.
32. **Koyama, Y.,** Structures and functions of carotenoids in photosynthetic systems, *J. Photochem. Photobiol. B,* 9, 265, 1991.
33. **Truscott, T. G.,** The photophysics and photochemistry of the carotenoids, *J. Photochem. Photobiol. B,* 6, 359, 1990.
34. **Siefermann-Harms, D.,** The light-harvesting and protective functions of carotenoids in photosynthetic membranes, *Physiol. Plant.*, 69, 561, 1987.
35. **Cogdell, R. J. and Frank, H.,** How carotenoids function in photosynthetic bacteria, *Biochim. Biophys. Acta,* 895, 63, 1988.
36. **Frank, H., Violette, C. A., Trautman, J. K., Shreve, A. P., Owens, T. G., and Albrecht, A. C.,** Photosynthetic carotenoids: structure and photochemistry, *Pure Appl. Chem.*, 63, 109, 1991.
37. **Gust, D., Moore, T. A., Benasson, R. V., Mathis, P., Land, E. J., Chachaty, C., Moore, A. L., Liddell, P. A., and Nemeth, G. A.,** Stereodynamics of intramolecular triplet energy transfer in carotenoporphyrins, *J. Am. Chem. Soc.*, 107, 3631, 1985.
38. **Siefermann-Harms, D. and Ninnemann, H.,** Pigment organisation in the light-harvesting chlorophyll a/b protein complex of lettuce chloroplasts: evidence obtained from protection of chlorophylls against proton attack and energy transfer, *Photochem. Photobiol.*, 35, 719, 1982.
39. **Krinsky, N. I.,** Non-photosynthetic functions of carotenoids, *Phil. Trans. R. Soc. Lond. B.*, 284, 581, 1978.
40. **Miki, W.,** Biological functions and activities of animal carotenoids, *Pure Appl. Chem.*, 63, 141, 1991.
41. **Farmilo, A. and Wilkinson, F.,** On the mechanisms of quenching of singlet oxygen in solution, *Photochem. Photobiol.*, 18, 447, 1983.
42. **Mathis, P. and Kleo, J.,** The triplet state of β-carotene and of analog polyenes of different length, *Photochem. Photobiol.*, 18, 343, 1973.
43. **Mathews-Roth, M. M., Wilson, T., Fujimore, E., and Krinsky, N. I.,** Carotenoid chromophore length and protection against photosensitization, *Photochem. Photobiol.*, 19, 217, 1974.
44. **Mathews-Roth, M. M. and Krinsky, N. I.,** Failure of conjugated octaene carotenoids to protect a mutant of *Sarcina lutea, Photochem. Photobiol.*, 11, 555, 1970.
45. **Di Mascio, P., Kaiser, S., and Sies, H.,** Lycopene as the most efficient biological carotenoid singlet oxygen quencher, *Arch. Biochem. Biophys.*, 274, 532, 1989.
46. **Murasecco-Suardi, P., Oliveros, E., and Braun, A. M.,** Singlet-oxygen quenching by carotenoids: steady state luminescence experiments, *Helv. Chim. Acta,* 71, 1005, 1988.
47. **Rodgers, M. A. J. and Bates, A. L.,** Kinetic and spectroscopic features of some carotenoid triplet states: sensitization by singlet oxygen, *Photochem. Photobiol.*, 31, 533, 1980.
48. **Burton, G. W. and Ingold, K. U.,** β-Carotene: an unusual type of antioxidant, *Science,* 224, 569, 1984.
49. **Burton, G. W.,** Antioxidant properties of carotenoids, *J. Nutr.*, 119, 109, 1990.

50. **Borland, C. F., Codgell, R. F., Land, E. J., and Truscott, T. G.,** Bacteriochlorophyll a triplet state and its interactions with bacterial carotenoids and oxygen, *J. Photochem. Photobiol.*, 3, 237, 1989.
51. **Moore, T. A., Gust, D., and Moore, A. L.,** The function of carotenoid pigments in photosynthesis and their possible involvement in the evolution of plants, in *Carotenoids: Chemistry and Biology*, Krinsky, N. I., Mathews-Roth, M. M., and Taylor, R. F., Eds., Plenum Press, New York, 1990, 223.
52. **Asada, K., Takahashi, M., Nakano, Y., and Shimada, S.,** Active oxygen in chloroplasts: suppression of singlet oxygen formation and scavenging of hydrogen peroxide, in *Oxygenases and Oxygen Metabolism*, Academic Press, New York, 1982, 255.
53. **Koyama, Y.,** Natural selection of carotenoid configurations by the reaction centre and light-harvesting complex of photosynthetic bacteria, in *Carotenoids: Chemistry and Biology*, Krinsky, N. I., Mathews-Roth, M. M., and Taylor, R. F., Eds., Plenum Press, New York, 1990, 107.
54. **Ashikawa, I., Miyata, A., Koike, H., Inoue, Y., and Koyama, Y.,** Light-induced structural change of β-carotene in thylakoid membranes, *Biochemistry*, 25, 6154, 1986.
55. **Young, A. J. and Britton, G.,** Carotenoids and stress, in *Stress Responses in Plants: Adaptation and Acclimation Mechanisms*, Alscher, R., Ed., Wiley-Liss, New York, 1990, 87.
56. **Satoh, K.,** Mechanisms of photoinactivation in photosynthetic systems. II. The occurrence and properties of two different types of photoinactivation, *Plant Physiol.*, 11, 29, 1970.
57. **Barry, P., Young, A. J., and Britton, G.,** Photodestruction of pigments in higher plants by herbicide action. I. The effect of DCMU (Diuron) on isolated chloroplasts, *J. Exp. Bot.*, 41, 123, 1990.
58. **Damm, I., Knoetzel, J., and Grimme, L. H.,** On the protective role of carotenoids in the PS I reaction centre and LHC I complexes of the thylakoid membrane, in *Progress in Photosynthesis Research*, Vol. II., Biggins, J. W., Ed., Kluwer Academic Press, Dordrecht, The Netherlands, 1987.
59. **Miller, N. and Carpentier, R.,** Energy dissipation and photoprotection mechanisms during chlorophyll photobleaching in thylakoid membranes, *Photochem. Photobiol.*, 54, 465, 1991.
60. **Ashikawa, I., Ito, M., Satoh, K., Koike, H., Inoue, Y., Saiki, T., Tsukida, K., and Koyama, Y.,** All-trans β-carotene-5,6-epoxide in thylakoid membranes, *Photochem. Photobiol.*, 46, 269, 1987.
61. **Young, A. J., Barry, P., and Britton, G.,** The occurrence of β-carotene-5,6-epoxide in the photosynthetic apparatus of higher plants, *Z. Naturforsch.*, 44c, 959, 1989.
62. **Ridley, S.,** Carotenoids and herbicide action, in *Carotenoid Chemistry and Biochemistry*, Britton, G. and Goodwin, T. W., Eds., Pergamon Press, Oxford, 1982, 353.
63. **Britton, G., Barry, P., and Young, A. J.,** Carotenoids and chlorophylls: herbicidal inhibition of pigment biosynthesis, in *Herbicides and Plant Metabolism*, Dodge, A. D., Ed., Cambridge University Press, Cambridge, 1989, 51.
64. **Bramley, P. M.,** Inhibition of carotenoid biosynthesis, in *Target Sites for Herbicides*, Kirkwood, R. C., Ed., Plenum Press, New York, 1991, 95.
65. **Young, A. J.,** Inhibition of carotenoid biosynthesis, in *Topics in Photosynthesis, Vol. 10, Baker, N. R. and Percival, M. P., Eds.*, Elsevier, Amsterdam, 1991, 130.
66. **Plummley, F. G. and Schmidt, G. W.,** Reconstitution of chlorophyll a/b light-harvesting complexes: xanthophyll-dependent assembly and energy transfer, *Proc. Natl. Acad. Sci. U.S.A.*, 84, 146, 1987.
67. **Lichtenthaler, H. K.,** Chloroplast biogenesis, its inhibition and modification by new herbicide compounds, *Z. Naturforsch.*, 39c, 492, 1984.
68. **Bartles, P. G. and Watson, C. W.,** Inhibition of carotenoid synthesis by fluoridone and norflurazon, *Weed Sci.*, 26, 198, 1978.
69. **Feierabend, J.,** Comparison of the action of bleaching herbicides, *Z. Naturforsch.*, 39c, 450, 1984.
70. **Feierabend, J., Schulz, U., Kemmerich, P., and Lowitz, T.,** On the action of chlorosis-inducing herbicides in leaves, *Z. Naturforsch.*, 34c, 1036, 1979.
71. **Feierabend, J. and Winkelhusener, T.,** Nature of photooxidative events in leaves treated with chlorosis-inducing herbicides, *Plant Physiol.*, 70, 1277, 1982.

72. **Burns, E. R., Buchanan, G. A., and Carter, M. C.,** Inhibition of carotenoid synthesis as a mechanism of action of amitrole dichlormate and pyricolor, *Plant Physiol.*, 47, 144, 1971.
73. **Young, A. J. and Britton, G.,** unpublished data.
74. **Abadia, A., Lemoine, Y., Tremoliers, A., Ambard-Bretteville, F., and Remy, R.,** Iron deficiency in pea: effects on pigment lipid and pigment-protein complex composition of thylakoids, *Plant Physiol. Biochem.*, 27, 679, 1989.
75. **Elstner, E. F., Oswald, W., and Youngman, R. J.,** Basic mechanisms of pigment bleaching and loss of structural resistance in spruce (*Picea abies*) needles: advances in phytomedical diagnostics, *Experientia*, 41, 591, 1985.
76. **Lange, O. L., Zellner, H., Gebel, J., Schramel, P., Kostner, B., and Czygan, F.-C.,** Photosynthetic capacity, chloroplast pigments and mineral content of the previous year's spruce needles with and without the new flush: analysis of the forest-decline phenomenon of needle bleaching, *Oceologia*, 73, 361, 1987.
77. **Asada, K. and Takahashi, M.,** Production and scavenging of active oxygen in photosynthesis, in *Photoinhibition*, Kyle, D. J., Osmond, C. B., and Arntzen, C. J., Eds., Elsevier, Amsterdam, 1987, 227.
78. **Price, A., Young, A. J., Beckett, P., Britton, G., and Lea, P. J.,** The effect of ozone on plant pigments, in *Current Research in Photosynthesis*, Vol. IV., Baltscheffsky, M., Ed., Kluwer Academic Press, Dordrecht, The Netherlands, 1990, 595.
79. **Wolfenden, J., Robinson, D. C., Cape, J. N., Paterson, I. S., Francis, B. J., Melhorn, H., and Wellburn, A. R.,** Use of carotenoid ratios, ethylene emissions and buffer capacities for the early diagnosis of forest decline, *New Phytol.*, 109, 75, 1988.
80. **Senser, M., Kloos, M., and Lutz, C.,** Influence of soil substrate and ozone plus acid mist on the pigment content and composition of needles from young spruce trees, *Env. Pollut.*, 64, 295, 1990.
81. **Young, A. J., Britton, G., and Senser, M.,** Carotenoid composition of needles of *Picea abies* L. showing signs of photodamage, *Z. Naturforsch.*, 45c, 1111, 1990.
82. **Young, A. J. and Putwain, P.,** unpublished data.
83. **Knox, J. P. and Dodge, A. D.,** Singlet oxygen and plants, *Phytochem.*, 24, 889, 1985.
84. **Knox, J. P. and Dodge, A. D.,** Photodynamic damage to plant leaf tissue by rose bengal, *Plant Sci. Lett.*, 37, 3, 1984.
85. **Young, A. J. and Britton, G.,** Carotenoids and oxidative stress, in *Current Research in Photosynthesis*, Vol. IV, Baltscheffsky, M., Ed., Kluwer Academic Press, Dordrecht, The Netherlands, 1990, 587.
86. **Barry, P., Evershed, R. P., Young, A. J., Prescott, M. C., and Britton, G.,** Characterization of carotenoid acyl esters produced in drought-stressed barley seedlings, *Phytochemistry*, in press.
87. **Young, A. J., Wellings, R., and Britton, G.,** The fate of chloroplast pigments during senescence of primary leaves of *Hordeum vulgare* and *Avena sativum*, *J. Plant Phys.*, 137, 701, 1991.
88. **Lichtenthaler, H. K.,** Die plastoglobuli von spinat; ihre grobe, isolierung und lipochinonzusammensetzung, *Protoplasma*, 68, 65, 1969.
89. **Trevini, M. and Steinmuller, D.,** Composition and function of plastoglobuli. II Lipid composition of leaves and plastoglobuli during beech leaf senescence. *Planta*, 163, 91, 1985.
90. **Thayer, S. S. and Bjorkman, O.,** Leaf xanthophyll content and composition in sun and shade determined by HPLC, *Photosyn. Res.*, 23, 331, 1990.
91. **Grill, D. and Pfeifhofer, W.,** Carotene in fichennadelen II. Quantitative untersuchungen, *Phyton*, 25, 1, 1985.
92. **Young, A. J. and Britton, G.,** The distribution of α-carotene in the photosynthetic pigment-protein complexes of higher plants, *Plant Sci.*, 64, 179, 1989.
93. **Young, A. J., Johnson, G., Rees, D., Noctor, G., and Horton, P.,** unpublished data.
94. **Schoner, S., Foyer, C., Lelandais, M., and Krause, G. H.,** Increase in activities of scavengers for active oxygen in spinach related to cold acclimation in excess light, in *Current Research in Photosynthesis*, Vol. II., Baltscheffsky, M., Ed., Kluwer Academic Press, Dordrecht, The Netherlands, 1990, 483.
95. **Lee S. and Britton, G.,** unpublished data.

Chapter 4

THE XANTHOPHYLL CYCLE

Barbara Demmig-Adams and William W. Adams III

TABLE OF CONTENTS

0-8493-6328-4/93/$0.00 + $.50

I. OVERVIEW

Exposure of plants to intense light can result in the formation of destructive activated oxygen species within the photosynthetic apparatus of leaves. (See Asada and Takahashi.[1]) Among these are (Figure 1B) the superoxide radical and its derivates that are formed when oxygen acts as the terminal electron acceptor of the photosynthetic electron transport chain, and singlet oxygen that is formed when oxygen interacts directly with the light-absorbing pigment system, i.e., with excited states of chlorophyll molecules that are unable to pass their energy on to other chlorophyll molecules and into the electron transport chain. In other words, such activated oxygen species are formed when there is an excess of light, i.e., more light is absorbed by leaves than can be utilized in photosynthetic carbon metabolism. It is important to note, however, that an absolute level of light that is nonexcessive for one plant may represent excess light for another plant (see below), or for that same plant under conditions which result in diminished rates of photosynthesis (e.g., at chilling temperatures or under water stress).

Over a range of photon flux densities (PFD) from darkness to low light, light limits photosynthesis, and is, therefore, nonexcessive (Figures 1A and 2). As PFD continues to increase, however, photosynthesis begins to saturate with respect to the photons being absorbed by chlorophyll, and light becomes excessive. The PFD at which the transition from limiting to excess light occurs depends strongly on the species and the growth and/or environmental conditions. Leaves with high maximum rates of photosynthesis typically show this transition at higher PFDs than do leaves with lower maximum rates of photosynthesis. Figure 3 shows light-response curves of photosynthesis and of the ratio of PFD to photosynthesis (a measure of the degree of excess light) for a sunflower leaf with a high maximum rate of photosynthesis and an ivy leaf with a low maximum rate of photosynthesis. The degree of excess light experienced by these two leaves at any given PFD is quite different. Furthermore, environmental stresses that depress the rate of photosynthesis can increase the amount of excess light that is absorbed by a leaf, even at a constant incident PFD. Such environmental stresses can include low temperatures, high temperatures, water stress, salinity stress, nutrient stress, and other stresses. (See Powles[2] and Demmig-Adams and Adams.[3])

The absorption of excess light can lead to an accumulation of excitation energy in the photochemical apparatus. Subsequently, this energy may be dissipated through the formation of the activated oxygen species. Several protective mechanisms may be activated to prevent or ameliorate this accumulation of excitation energy. One major photoprotective process that occurs ubiquitously in higher plants is the thermal dissipation of excess excitation energy directly within the photochemical apparatus (Figure 1C). This thermal dissipation of energy counteracts and potentially prevents the accumulation of excess excitation energy in the photosynthetic apparatus. This energy dissipation process, involving the xanthophyll zeaxanthin,[4,5] will be the focus of this chapter.

Through its involvement in this thermal energy dissipation process, zeaxanthin has a special role among all of the carotenoids in the protection against an excess of light. Previously, the role of carotenoids in photoprotection had been thought to

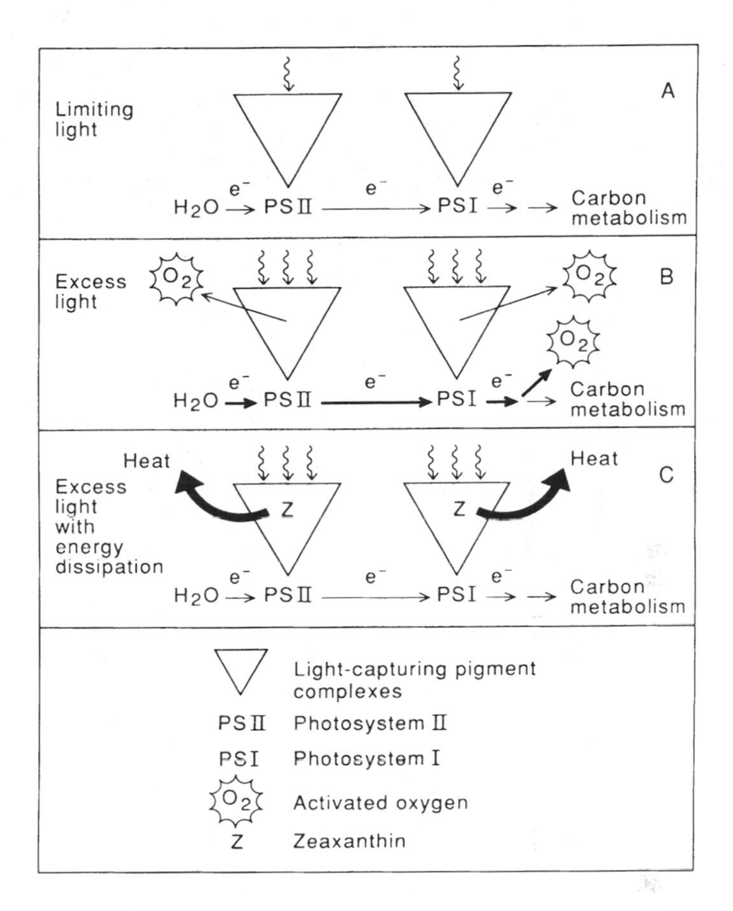

FIGURE 1. Schematic depiction of the energy flow through the photosynthetic electron transport chain under conditions of limiting and excess light. Part A describes the light-capturing and electron transport system under conditions when light is not excessive and no accumulation of excitation energy occurs. Part B describes the situation when excess light leads to an accumulation of excitation energy within the photochemical systems resulting in the formation of activated oxygen species, e.g., in the light-capturing pigment system (singlet oxygen, $^1O_2$) and at the end of the electron transport chain through reduction of O_2 by PS I (to superoxide, O_2^-). Part C describes the situation when thermal energy dissipation (associated with the xanthophyll zeaxanthin) counteracts the accumulation of excitation energy under excess light. As long as the capacity of energy dissipation is sufficient to remove all excess excitation energy, no accumulation occurs, and no activated-oxygen species are presumably formed within the system.

be through either a direct interaction with singlet oxygen, leading to its de-excitation, or through an interaction with the triplet-excited state of chlorophyll, that species of chlorophyll that gives rise to singlet oxygen.[6,7] The thermal energy dissipation process that involves zeaxanthin is unique. It is a novel mechanism, diverting energy, presumably before the triplet state of chlorophyll is ever formed (see below).

A variety of protective mechanisms function to prevent the accumulation of excitation energy in the photosynthetic apparatus. Mechanisms that prevent the

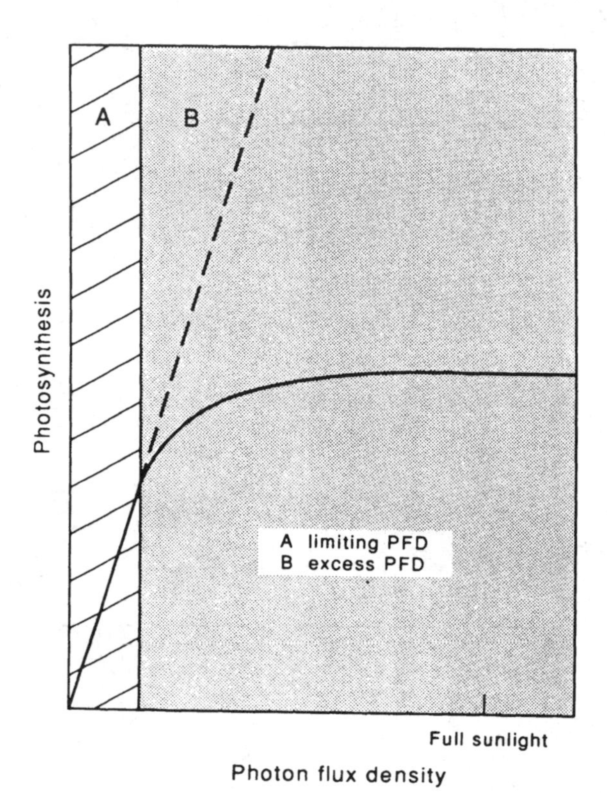

FIGURE 2. Typical light-response curve of the photosynthetic rate of a leaf. The response is characterized by an initial linear increase in photosynthesis with increasing photon flux density, followed by a region where further increases in PFD no longer influence photosynthesis, since the maximal capacity of carbon metabolism has been reached. Over this range of high PFDs, light is thus absorbed in excess of the amount that can be utilized in photosynthetic carbon metabolism, and can potentially lead to an accumulation of excitation energy within the photochemical system.

absorption of excessive light by the light-harvesting pigment system will not be discussed in detail, but include: a decreased absorbance of leaves, alteration in characteristics of the leaf surface that decrease transmittance of light to the chlorophyll-containing cells,[8] leaf movements,[9] and chloroplast movements.[10,11] Among all of these preventive mechanisms, however, the zeaxanthin-associated thermal energy dissipation process appears to be the one that is ubiquitous and the most flexible. The other mechanisms operate on a slower time scale than does the zeaxanthin-associated process, and/or are restricted to certain groups of species.

The zeaxanthin content of leaves is regulated, such that the accumulation of zeaxanthin is strictly correlated with the level of excess light.[4] Zeaxanthin is formed in the thylakoid membranes of leaves in a rapid, light-dependent reaction in the xanthophyll cycle.[12-14] These interconversions of xanthophylls are the only short-term changes in the carotenoid composition of the photosynthetic apparatus.[15] They occur in response to changes in the balance between the absorption of light by

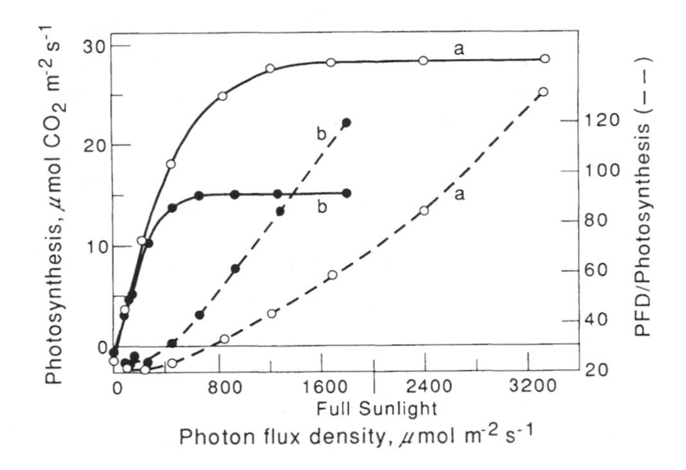

FIGURE 3. Light response of photosynthetic rates and the ratio of PFD to photosynthesis for two leaves with different maximal capacities of photosynthetic carbon metabolism. One is a leaf (a) of a sunflower (*Helianthus annuus*) plant that developed in full sunlight; the other (b) a leaf of an English ivy (*Hedera helix*) plant which developed under diffuse (scattered from the sky) light. The ratio of PFD to photosynthesis is a measure of the degree of excess light absorbed by each leaf. Photosynthesis was measured as CO_2 exchange in ambient air. (Data from Demmig-Adams, B., Winter, K., Krüger, A., and Czygan, F.-C., *Plant Physiol.*, 90, 881, 1989.)

chlorophyll and the utilization of light in photosynthetic carbon metabolism. The formation and removal of zeaxanthin, i.e., the turnover of the xanthophyll cycle, is observed under natural conditions during the course of the day in sun-exposed habitats. The total size of the xanthophyll cycle pool (i.e., the sum of the components violaxanthin + antheraxanthin + zeaxanthin), and thus the capacity for zeaxanthin formation, undergoes pronounced acclimation to the light environment, and increases with increasing degrees of excess light. Thus, the capacity of leaves for zeaxanthin formation is regulated in response to the demand for zeaxanthin-associated energy dissipation.

This chapter will, therefore, focus on the xanthophyll cycle pigments as three of the five ubiquitous xanthophylls of green leaves. The other two xanthophylls, lutein and neoxanthin, do not exhibit consistent changes in their levels in response to the light environment;[15,16] neither is there other evidence to suggest a specific function of these particular xanthophylls under high light stress.

II. BIOSYNTHESIS AND COMPARTMENTATION

Among the xanthophylls typically found in the leaves of higher plants, only lutein is a derivative of α-carotene (β,ε-carotene). All other major xanthophylls are derived from β-carotene (β,β-carotene); zeaxanthin, antheraxanthin, violaxanthin, and neoxanthin. A portion of the presumed biosynthetic pathway[17] of the formation of the xanthophyll cycle components is depicted in Figures 4 and 5. Neoxanthin is not included in this diagram, but it has been suggested that neoxanthin can be

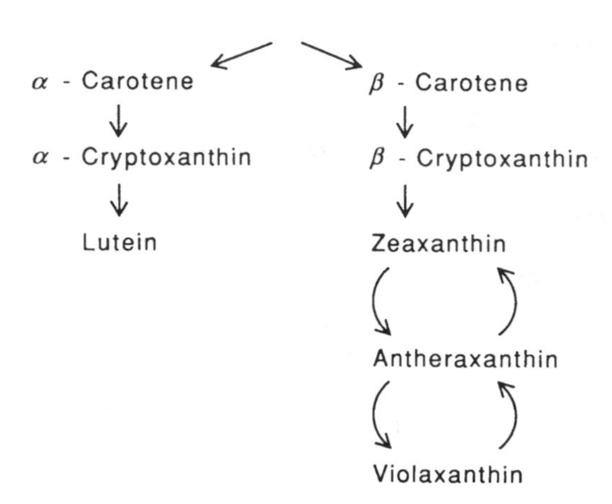

FIGURE 4. Presumed final steps of the pathway of xanthophyll biosynthesis. Shown are α-carotene (β,ε-carotene) and its derivates, such as lutein (dihydroxy-β,ε-carotene), as well as β-carotene (β,β-carotene) and its derivates, zeaxanthin, antheraxanthin, and violaxanthin. Neoxanthin, that is also a derivate of β-carotene, is not shown in this diagram.

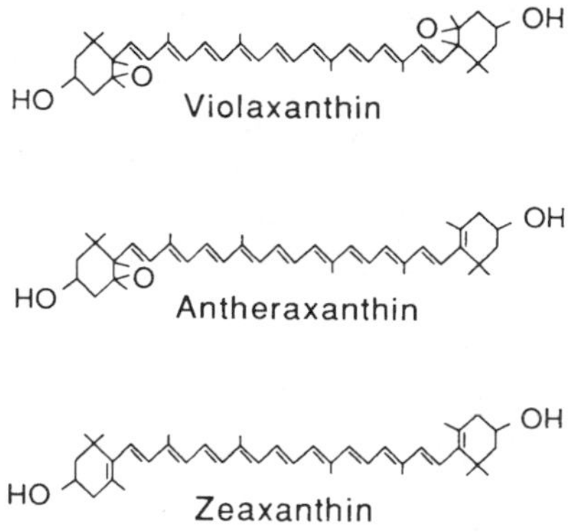

OH
O
O
HO
Violaxanthin

OH
O
HO
Antheraxanthin

OH
HO
Zeaxanthin

FIGURE 5. Structures of the three xanthophylls of the xanthophyll cycle: violaxanthin (diepoxide; diepoxyzeaxanthin), antheraxanthin (monoepoxide; monoepoxyzeaxanthin), and zeaxanthin (epoxide-free; dihydroxy-β,β-carotene).

formed from violaxanthin (see below). Excess light specifically stimulates the β,β-carotenoid pathway, and leads to the accumulation of large amounts of β-carotene, and, particularly, the components of the xanthophyll cycle, zeaxanthin, antheraxanthin, and violaxanthin.[15,16] The xanthophyll cycle appears to be present throughout all families of higher plants.[15,16] From indirect evidence, it seems that, among the three xanthophyll cycle components, zeaxanthin is formed first through hydroxylation of β-carotene. Antheraxanthin and violaxanthin are formed subsequently through epoxidation of zeaxanthin. The epoxidation state of the xanthophyll cycle is thereafter regulated by light. It has recently been reported that a mutant of

Arabidopsis, unable to form more than trace amounts of violaxanthin, accumulated larger amounts of both β-carotene and zeaxanthin than did the wild type.[18,19] This mutant also possessed reduced levels of lutein as a derivative of α-carotene,[19] and was therefore enhanced in the β,β-carotenoid pathway, apparently at the expense of the β,ε-carotenoid pathway. This mutant contained only trace amounts of neoxanthin as well, a result consistent with a formation of neoxanthin from violaxanthin.[18,19]

Under excessive light, zeaxanthin can be formed rapidly from violaxanthin (and antheraxanthin, the intermediate in the cycle) through an enzymatic de-epoxidation reaction. It has also been suggested that additional zeaxanthin can be synthesized relatively rapidly from β-carotene when light is excessive.[20,21] More information, however, on the biosynthetic pathway, and in particular on its regulation through high/excess light, is needed.

The available information on the compartmentation of the various leaf xanthophylls is also scarce. In green leaves, xanthophylls generally accumulate in the chloroplasts, and within the chloroplasts, they are integral to the thylakoid membrane.[22] Their precise localization within the thylakoid membrane has not been established. It has, however, been shown in an elegant series of experiments that the xanthophyll cycle occurs within the supracomplexes of both Photosystem II (PS II) and Photosystem I (PS I)[57] (see also Demmig-Adams and Adams[5]). The three components of the xanthophyll cycle are associated with both photosystems, and undergo de-epoxidation and epoxidation to a similar extent in both systems. Within each photosystem, it is unclear with which components the xanthophyll cycle is associated. A portion of the xanthophyll cycle components may be bound to the peripheral light-harvesting complexes, e.g., light-harvesting complex-II (LHC-II), but components of the cycle may also be present in other chlorophyll-protein complexes.[22]

The final steps of the biosynthesis of the xanthophyll cycle components have been proposed to be associated with the chloroplast envelope.[23] These include the synthesis of zeaxanthin from β-carotene, followed by the epoxidation of zeaxanthin to antheraxanthin and subsequently to violaxanthin (Figure 4). Zeaxanthin-epoxidase activity may occur in chloroplast envelope membranes.[23] De-epoxidation of violaxanthin to antheraxanthin and zeaxanthin has, however, been observed exclusively within thylakoid membranes; no de-epoxidase activity has been reported for the envelope membranes. It has been postulated that violaxanthin can be transferred from the envelope to the thylakoids, where the entire xanthophyll cycle takes place.[24]

III. FUNCTION AND METABOLISM

A. THERMAL ENERGY DISSIPATION

Thermal dissipation of energy in the photochemical apparatus is a regulated process. Dissipation is not usually observed under limiting light (Figure 6). Under excess light, the dissipation activity increases with further increases in PFD. This increase in thermal dissipation occurs independent of whether the excess energy results from increases in PFD (as shown in Figure 6) or from stress induced decreases

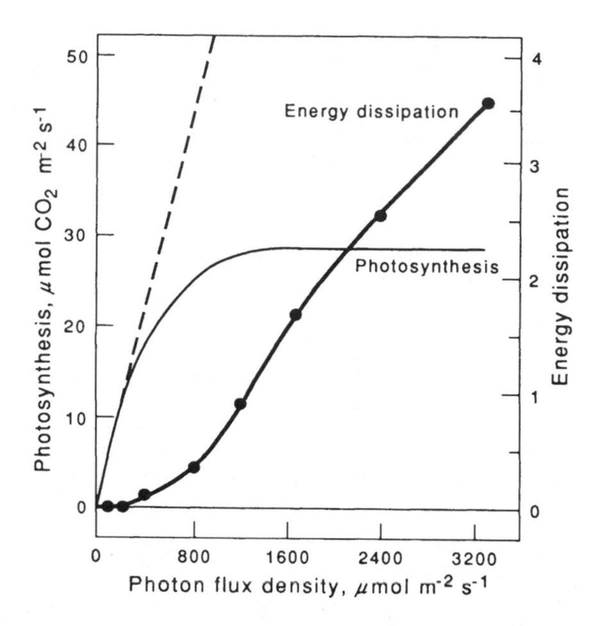

FIGURE 6. Light response of photosynthesis and thermal energy dissipation activity for a sunflower leaf developed in full sunlight. Photosynthesis was measured as CO_2 exchange (cf. Figure 3). Energy-dissipation activity was quantified from the lowering of the yield of maximum chlorophyll fluorescence, and is expressed as the ratio of (unquenched fluorescence in darkness to quenched fluorescence at each PFD) minus 1 (Stern-Volmer quenching). (See Demmig-Adams[4] and Bilger and Börkman.[33]) Chlorophyll fluorescence was determined at ambient temperature with the pulse-saturation method,[56] where maximum fluorescence can be determined both in the dark (unquenched state) and under ongoing illumination (quenched state). (Data from Demmig-Adams, B., Winter, K., Krüger, A., and Czygan, F.-C., *Plant Physiol.*, 90, 881, 1989.)

in the rate of photosynthesis in the field (see below) without any change in the incident PFD.

When excitation energy is dissipated thermally in the photochemical system, the operation of this process results in a decrease in the efficiency of photosynthetic energy conversion.[25,26] This effect accounts for the fact that this energy dissipation process was initially misinterpreted as damage to the photochemical system. Damage to the photosynthetic apparatus can indeed cause a loss of photosynthetic activity.[27] However, it has now been clearly shown that, under a variety of environmental conditions,[20,28-31] a low efficiency of photosynthetic energy conversion can result from ongoing thermal-energy dissipation that removes excitation energy before it reaches the reaction centers, and thereby competes with the process of photosynthesis for excitation when the PFD is rate-limiting for photosynthesis. At high PFDs, the effect of this decrease in photosynthetic efficiency can be negligible, but not at limiting PFDs.

In the absence of any additional environmental stress, the activity of thermal energy dissipation decreases rapidly upon return from excess to low PFDs, and, therefore, only rapidly reversible decreases in the rate of photosynthesis are observed upon return to a low PFD.[4] However, clearly there are also situations when thermal energy dissipation in excess PFD is not rapidly reversible upon return to low PFD. This has been particularly observed when plants have been subjected to long-term water stress[20] or salinity stress.[29] Under these conditions, thermal energy dissipation activity remains high throughout the entire day (with excess light) and night cycle. In such plants, this high level of thermal-energy dissipation that is thought to protect the photosynthetic system from damage is, therefore, accompanied by a probably small cost: a lowered photosynthetic rate under limiting light for some time upon return to favorable conditions.

Thermal energy dissipation in the photochemical system is currently identified and quantified through its effect on the yield of fluorescence emitted by chlorophyll (see Demmig-Adams,[4] Demmig-Adams and Adams,[5] and Krause and Weis[32]). Excitation energy can be dissipated via several pathways: through photochemistry, or, alternatively, through thermal energy dissipation, and also, to a very small extent, through chlorophyll fluorescence. The emission of fluorescence is high when photosynthesis rates and/or thermal energy dissipation is/are low or zero, and low when photosynthesis rates and/or thermal energy dissipation is/are high. It is possible to exclude the effect of photosynthesis and use the yield of fluorescence as a quantitative indicator of the activity of the thermal energy dissipation process.[4]

When thermal energy dissipation in leaves is prevented, several adverse effects can often be observed. The reduction state of PS II, which is a measure of the balance between the rate at which light energy is captured and the rate at which it is consumed in the biochemistry of photosynthesis, is markedly increased due to an increased accumulation of excitation energy within the light-harvesting system.[33,34] As a consequence, photosynthetic systems illuminated with excess light in the absence of thermal energy dissipation can suffer an irreversible loss of photosynthetic activity that is caused by damage to the photosynthetic system.[33,35]

B. ROLE OF THE XANTHOPHYLL CYCLE IN ENERGY DISSIPATION

1. Characteristics of the Xanthophyll Cycle

The xanthophyll cycle consists of a cyclic sequence of two reactions, catalyzed by two different enzymes (Figure 7).[12-14] The two sequences involve the de-epoxidation of violaxanthin (di-epoxide) to antheraxanthin (mono-epoxide) to zeaxanthin (epoxide-free), catalyzed by a de-epoxidase, and an epoxidation sequence reversing these two steps that is catalyzed by an epoxidase. *In vitro,* these reactions can be driven in either direction in darkness, although *in vivo,* the de-epoxidation reaction depends upon excess light, and the epoxidation reaction is stimulated by low light.

The xanthophyll cycle provides a means of rapidly interconverting (within minutes) the three xanthophylls of the cycle, in response to changes in the balance between the absorption of light by chlorophyll and the utilization of light through photosynthetic carbon metabolism. The regulation of the cycle that is exercised by light results in an accumulation of the de-epoxidized carotenoids (particularly zeaxanthin) under excess light, whereas the epoxidized forms (particularly violaxanthin) are accumulated under limiting light. Thus, in limiting light, violaxanthin is the predominant xanthophyll cycle component. Increasing amounts of zeaxanthin are

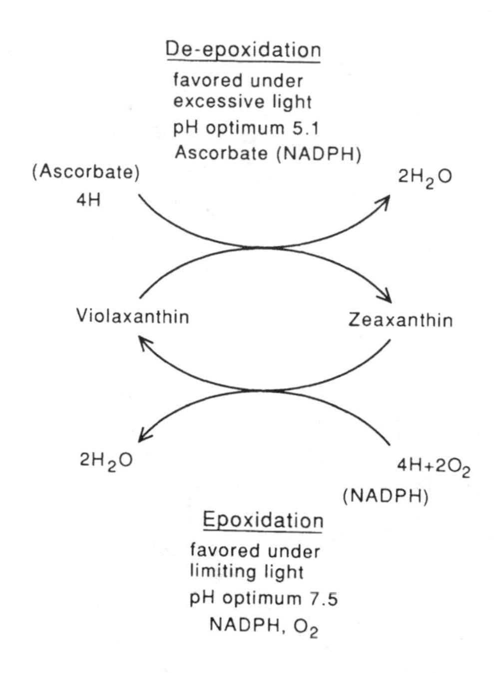

FIGURE 7. Schematic summary of the factors involved in the regulation of the xanthophyll cycle. Shown is only the sum of the two-step reaction from violaxanthin (via antheraxanthin) to zeaxanthin and the two-step reaction in the opposite direction. The pH optima shown are those determined for the isolated de-epoxidase enzyme and for the epoxidation reaction.

formed as incident PFD increases beyond that required to saturate photosynthesis (Figure 8). The transition point between light-limited and light-saturated photosynthesis is, therefore, also the transition point between the predominance of violaxanthin, i.e., a high epoxidation state of the xanthophyll cycle, and beginning de-epoxidation to form increasing amounts of zeaxanthin. The levels of the intermediate, antheraxanthin, increase initially with increasing PFDs, but fall again at even higher PFDs. This light regulation of the cycle, i.e., the coupling of the epoxidation state to the degree to which light is in excess, is achieved through the properties of the two enzymes involved in the cycle, as well as through variation in the co-substrate concentrations.

As summarized in Figure 7, the de-epoxidase has a low pH optimum, whereas the epoxidation reaction has a high pH optimum. Since both enzymes can apparently be active simultaneously in the light when the pH of the thylakoid lumen is low whereas the pH of the chloroplast stroma is high, it has been proposed that the de-epoxidase is located at the interface between the thylakoid membrane and the inner lumen.[14] The epoxidase is possibly buried deep within the membrane, away from the lumen side.[14] This model still awaits experimental confirmation.

NADPH, and thus photosynthetic electron transport, supplies reducing equivalents for the reductive de-epoxidation as well as for the epoxidation reaction involving the incorporation of molecular oxygen (Figure 7). However, NADPH is

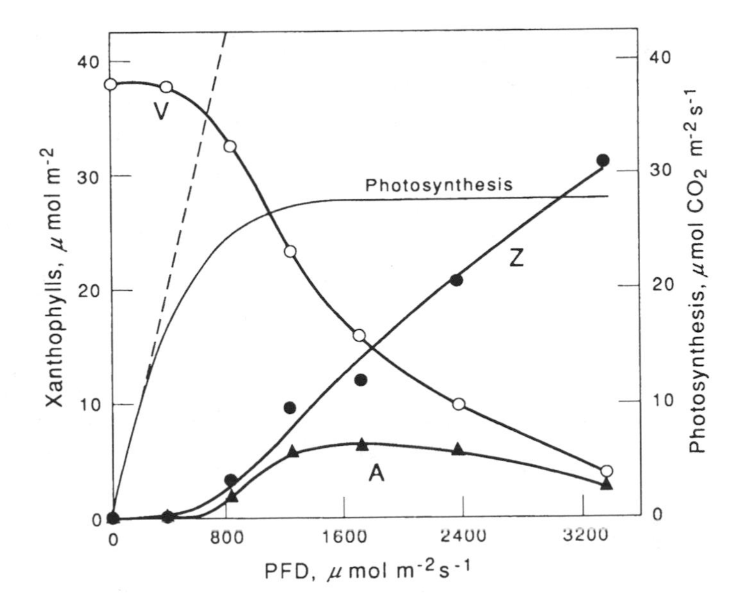

FIGURE 8. Light response of photosynthesis and the content of the three components of the xanthophyll cycle, violaxanthin (V), antheraxanthin (A), and zeaxanthin (Z), for a sunflower leaf that developed in full sunlight. The leaf was consecutively exposed to each photon flux density in air for 30 to 45 min, at which time a portion of the leaf was sampled for pigment composition. The pigment composition was determined by extraction of the samples in acetone and analysis by thin layer chromatography.[40,47] (Data from Demmig-Adams, B., Winter, K., Krüger, A., and Czygan, F.-C., *Plant Physiol.*, 90, 881, 1989.)

only involved directly as a co-substrate in the epoxidation reaction. The endogenous reductant for the de-epoxidation reaction is presumably reduced ascorbate which becomes rereduced through NADPH.[1] Hence, reduced ascorbate serves not only directly as an antioxidant, but also facilitates the formation of zeaxanthin that is, in turn, engaged in the prevention of the formation of activated oxygen species. This consumption of NADPH for the rereduction of ascorbate contributes to linear photosynthetic electron flow and the formation of a transthylakoid pH gradient.[36] (See also Schreiber, Reising, and Neubauer.[37]) A minimal transthylakoid pH gradient is, in turn, the prerequisite for the formation of zeaxanthin (see above), and has also been postulated to be necessary for the actual function of zeaxanthin in thermal energy dissipation.[34,38,39] Thus, multiple interactions relate ascorbate metabolism and xanthophyll cycle metabolism.

2. Evidence for an Involvement of Zeaxanthin in Energy Dissipation

A comparison of Figures 6 and 8 reveals that increased thermal energy dissipation occurs in response to excess PFD, and proceeds in a very similar fashion to the increase in the zeaxanthin level. Linear correlations between these two parameters have been reported for a range of plant species and under a variety of conditions.[20,34,38,40-44] Evidence for a direct involvement of zeaxanthin in thermal energy

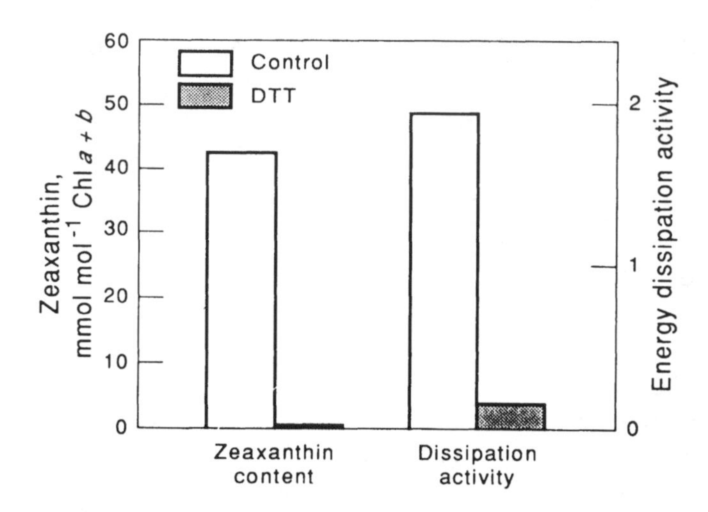

FIGURE 9. Response of zeaxanthin levels and thermal energy dissipation in spinach leaves treated with dithiothreitol. Leaves were collected from low light, the cut petioles placed in a solution of either water (control) or 5 m*M* DTT, and the leaves exposed to 40 μmol photons m^{-2} s^{-1} for 90 min. Subsequent to this pretreatment, the leaves were exposed to a PFD of 900 μmol photons m^{-2} s^{-1} for 10 min. The energy dissipation activity was quantified from the yield of chlorophyll fluorescence. (See legend of Figure 6 for further details.) A value for this energy-dissipation activity of 0.54 present under limiting light in both controls and DTT-treated leaves was assumed to be unrelated to energy dissipation, and was subtracted from both sets of values. (Data from Demmig-Adams, B. et al., *Plant Physiol.*, 92, 293, 1990.)

dissipation (see Demmig-Adams and Adams[3,5] and Demmig-Adams[4]) stems from studies with an inhibitor of zeaxanthin formation in the xanthophyll cycle, dithiothreitol (DTT).[33-35,38,44] Dithiothreitol inhibits the activity of the de-epoxidase,[45] and completely prevents zeaxanthin formation in leaves,[46] such that even at excess PFDs, levels of violaxanthin remain high, and levels of antheraxanthin and zeaxanthin remain at zero. Hence, DTT also prevents the zeaxanthin-associated energy dissipation process in leaves (Figure 9).[33-35] It has also recently been shown that in leaves that were allowed to accumulate zeaxanthin, followed by treatment with DTT, thermal energy dissipation occurred to the same extent as in leaves that were not treated with DTT.[58] These data indicate that the inhibition of thermal energy dissipation by DTT is through the inhibition of zeaxanthin formation.

Other studies have employed photosynthetic organisms that naturally lack the xanthophyll cycle.[47-50] Aerobic photosynthetic organisms that do not possess the xanthophyll cycle include the blue-green algae or cyanobacteria. It was shown that under excess light, blue-green algal lichens without zeaxanthin were unable to dissipate excitation energy. The accumulated excitation energy in the photochemical system resulted in sustained reductions in photosynthetic efficiency following exposure to excess light. In contrast, green algal lichens (which possess the xanthophyll cycle) developed strong thermal energy dissipation during exposure to high light, and did not experience any sustained changes in photosynthesis.[48,49] Similarly, blue-green algal lichens that contained zeaxanthin (presumably synthesized from β-carotene) developed strong energy dissipation activity during exposure to excess light, similar to that developed by the green algal lichens.[48]

In all of these studies, the presence of zeaxanthin was associated with the occurrence of a high thermal energy dissipation activity, whereas in the zeaxanthin-free systems, little or no thermal energy dissipation was observed. Furthermore, exposure to excess light led to an accumulation of excitation energy in the photochemical reaction centers in the zeaxanthin-free systems, and subsequent irreversible decreases in photosynthetic efficiency. Zeaxanthin-associated thermal energy dissipation is accompanied by a rapidly reversible decrease in photosynthesis rates, and the absence of this zeaxanthin-associated process leads to a qualitatively different effect that involves adverse effects to the photosynthetic apparatus. A role of the zeaxanthin-associated energy dissipation process in the photoprotection of the photosynthetic system is well documented.

The precise mechanism of how zeaxanthin functions in thermal energy dissipation has not yet been resolved. Zeaxanthin somehow facilitates the return of excited state chlorophyll to the ground state without leading to any adverse reactions. Since zeaxanthin-associated energy dissipation results in a lowered yield of chlorophyll fluorescence, it has been proposed that the excited singlet state of chlorophyll, the source of fluorescence, is returned to the ground state.[4] Zeaxanthin and the excited singlet state of chlorophyll could hypothetically interact directly through an unknown mechanism, (see Demmig-Adams[4]) or they could interact indirectly through an effect of zeaxanthin on the conformation or aggregation of chlorophyll-binding complexes.[4,5,14] (See also Ruban et al.[51]) In either case, the excitation energy would be converted to heat.

A role of zeaxanthin in the thermal dissipation of excess excitation energy is intriguing, since the presence of zeaxanthin in the photosynthetic membrane is restricted to conditions of excess light. The presence of a factor with an active role in energy dissipation would not be compatible with the optimization of photosynthetic activity under light-limited conditions. Zeaxanthin is the only carotenoid that undergoes such changes in response to excess or limiting light. It is not known which property of the zeaxanthin molecule allows this special function as opposed to the other xanthophylls or carotenes.

3. Environmental Influence on the Xanthophyll Cycle

a. Diurnal Response

Plants which grow in full sunlight commonly experience excess light during midday, when more light is absorbed by the leaves than can be utilized through photosynthetic carbon metabolism. Thus, major conversions of violaxanthin to zeaxanthin and back to violaxanthin can generally be observed in these plants over the course of the day in the field (Figure 10).[15,52,53] There is a remarkably close association between the changes in PFD incident on leaves and changes in their zeaxanthin level. Since the total pool size of zeaxanthin + antheraxanthin + violaxanthin typically remains constant over the course of a day, the changes in zeaxanthin content are quantitatively matched by changes in violaxanthin (and antheraxanthin). The close association between incident PFD and the zeaxanthin levels has been found in leaves with very different exposures. For example, east-

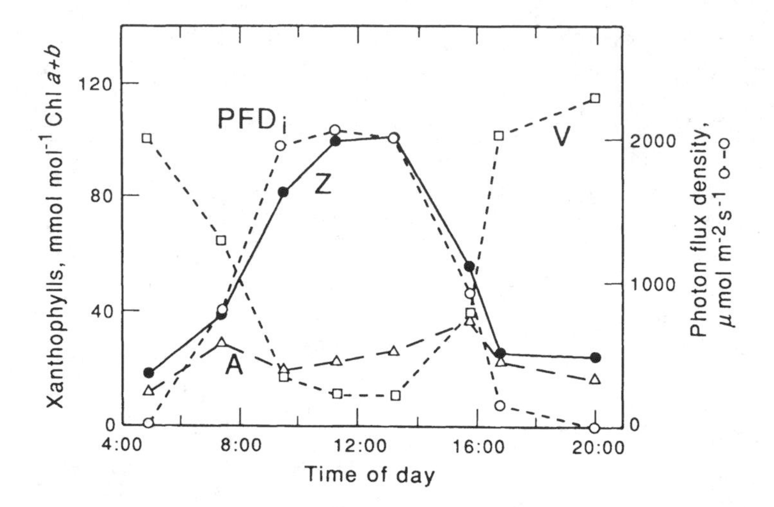

FIGURE 10. Changes in incident PFD and the levels of the three components of the xanthophyll cycle, violaxanthin, antheraxanthin, and zeaxanthin, over the course of a clear day for a leaf of the perennial shrub *Euonymus kiautschovicus*. The data[52] were collected from a sun-exposed (facing south) leaf of the shrub on September 15, 1990 in Colorado. The samples were collected rapidly and frozen in liquid nitrogen, followed by extraction and analysis by high performance liquid chromatography.[16,52] (Data from Adams, W. W. III and Demmig-Adams, B., *Planta,* 186, 390, 1992.)

facing leaves that receive peak irradiance in the morning also show peak zeaxanthin levels in the morning, and west-facing leaves receiving peak irradiance in the afternoon show peak zeaxanthin levels in the afternoon.[53] The degree to which violaxanthin is converted into zeaxanthin depends on the degree of excess light intercepted by the leaf, and therefore on the ratio of the photosynthetic rate to the absorbed PFD. In full sunlight, those leaves and plants with lower light-saturated rates of photosynthesis (that therefore experience a greater excess of light) form more zeaxanthin, and exhibit a more highly de-epoxidized state of the xanthophyll cycle than those with higher light-saturated rates of photosynthesis.[15,16,52] It is, therefore, possible to make predictions about the photosynthetic activity of a leaf at a known PFD from the epoxidation state of the xanthophyll cycle.[16]

b. Stress Response

Exposure of plants to various environmental stress factors almost inevitably leads to a decrease in the rate of photosynthesis. In turn, the incident light becomes more excessive, even at a constant absolute light level. In one study in which plants were subjected to progressively increasing water stress, it was found that increased amounts of zeaxanthin accumulated, and that this zeaxanthin was retained in the leaves overnight (Figure 11).[20] Sustained increases in thermal energy dissipation were also observed in these leaves. Similarly, increased levels of thermal energy dissipation (presumably associated with zeaxanthin) that were maintained during the night were observed in mangrove plants exposed to high levels of salinity.[29] More recently, we have found that chilling stress resulted in a retention of zeaxanthin

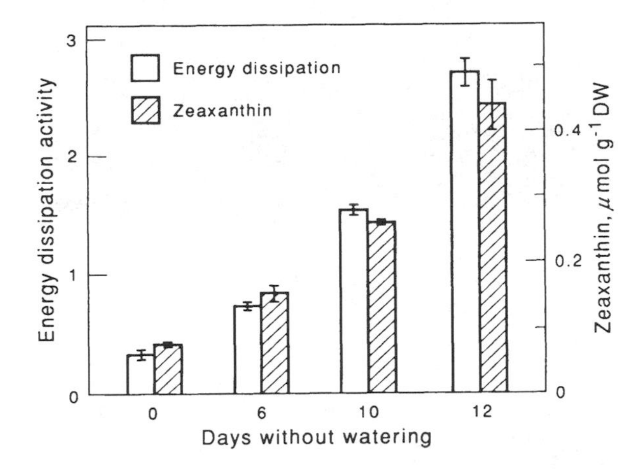

FIGURE 11. Effect of water stress on the zeaxanthin content and the thermal energy dissipation activity in leaves of the evergreen shrub *Nerium oleander*. Both the zeaxanthin content and the energy dissipation activity were determined at the end of the 12-h dark period. The energy dissipation activity was quantified from the decreased yield of chlorophyll fluorescence (cf. legend of Figure 6). However, in this case, the maximum yield of fluorescence was determined only in the dark, and not under illumination. This maximum yield was less on each consecutive morning (end of night), and these lowered values were used to calculate the ratio of unquenched to quenched fluorescence. The unquenched value was the value determined at the end of the night for well-watered leaves. There was no change in the incident PFD during this period (the leaves were restrained horizontally). Pigment analysis was performed by thin layer chromatography.[40,47] (Data from Demmig, B., Winter, K., Krüger, A., and Czygan, F.-C., *Plant Physiol.*, 87, 17, 1988.)

within leaves of several evergreen species throughout the day/night cycle under natural conditions.[59] Thus, plants which are stressed may simply retain the zeaxanthin rather than expend energy in the operation of the cycle each day.

Other stress factors, such as nitrogen limitation[54] and iron deficiency,[55] have also been reported to lead to an increased zeaxanthin accumulation. An accumulation of zeaxanthin under high light in combination with other stress factors presumably acts to counteract the increased potential for an accumulation of excitation energy in the photosynthetic apparatus under these conditions. It is not known to what extent the zeaxanthin-associated dissipation process can prevent or ameliorate this accumulation. Other antioxidant systems, particularly components of ascorbate and glutathione metabolism, also respond to environmental stresses. As indicated above, ascorbate metabolism is also linked directly to the xanthophyll cycle. Further investigation into these questions is needed, particularly simultaneous studies of metabolism of the xanthophyll cycle and other antioxidants under environmental stresses in the field.

c. *Response to Varying Light Environments*

The total content of the three components of the xanthophyll cycle varies in leaves acclimated to full sun vs. shade. The components of the xanthophyll cycle in leaves which have developed in deep shade typically constitute about 10% of

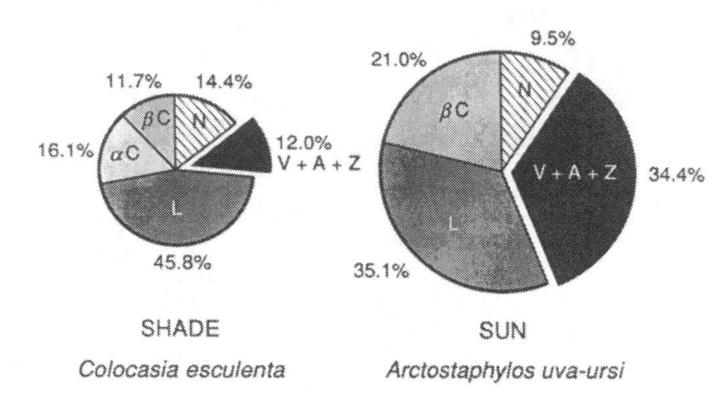

FIGURE 12. Contribution in percent of the various leaf carotenoids to the total carotenoid content of a shade-grown leaf of *Colocasia esculenta* and a sun-grown leaf of *Arctostaphylos uva-ursi*. The relative size of each pie diagram reflects the relative amounts of the total carotenoids present in each leaf (in millimole carotenoid mol^{-1} chlorophyll a + b). *Colocasia esculenta* is a rainforest species, and was growing under a dense canopy in the Denver Botanical Gardens. *Arctostaphylos uva-ursi* is an evergreen shrub growing on a south-facing slope in the front range of the Rocky Mountains in Colorado. Pigments were analyzed by thin layer chromatography (*A. uva-ursi*) or by high performance liquid chromatography (*C. esculenta*). (Data from Demmig-Adams, B. and Adams, W. W. III, *Plant Cell Environ.*, and Adams, W. W. III et al., *Oecologia*, 90, 404, 1992.)

the total carotenoids, whereas the components of the xanthophyll cycle in leaves which have developed in full sunlight may represent as much as 35% of the total carotenoids (Figure 12). Similar patterns have been observed among a large number of other shade plants (such as understory rainforest species) vs. sun plants (e.g., various crop species), as well as between sun and shade leaves of the same species.[15,16] (See also Demmig-Adams et al.[21]) Thus the acclimation in the pool size of the xanthophyll cycle is the most pronounced change observed in any carotenoid component in response to PFD. Only β-carotene, the precursor of the xanthophyll cycle components, can also be found in higher levels in leaves which develop in higher PFD. High or excess light, therefore, appears to stimulate specifically β,β-carotenoid biosynthesis (β-carotene and the components of the xanthophyll cycle).[15,16] This response is consistent with the role of the xanthophyll cycle in protecting the photosynthetic apparatus against damage from accumulated excitation energy. Larger pool sizes of violaxanthin + antheraxanthin + zeaxanthin in leaves exposed to full sunlight allow a greater potential for forming zeaxanthin, and, thus, a greater capacity for the zeaxanthin-associated dissipation of accumulated excitation energy.

Lutein and neoxanthin, the other two xanthophylls typically found in the leaves of higher plants, do not vary consistently in response to the light environment. It is, therefore, assumed that these carotenoids do not function in the acclimation of the photosynthetic apparatus to high or excess light.

It has recently been reported for a number of species (mostly rapidly growing mesophytes) that leaves growing in full sunlight and exhibiting high rates of photosynthesis also contained very large concentrations of xanthophyll cycle components. These plants utilized, on average, less than 50% of this pool to form zeaxanthin at peak irradiance at noon.[15] This response suggests that these plants, which were grown in full sunlight, have an ample capacity for zeaxanthin-associated energy dissipation. In contrast, other species (mostly more slowly growing perennial species) that were also growing in full sunlight possessed lower photosynthetic capacities, and converted nearly the entire xanthophyll cycle pool into zeaxanthin at peak irradiance. It remains to be investigated further how fully various plants acclimated to high light can prevent an accumulation of excitation energy, and thus the formation of activated oxygen species.

IV. SUMMARY

Recent evidence indicates that zeaxanthin is involved in a thermal energy dissipation process that acts to protect the photosynthetic apparatus against excess light. The evidence for an involvement of zeaxanthin in energy dissipation includes the inhibition of thermal energy dissipation by an inhibitor of the violaxanthin de-epoxidase, DTT, and the absence of thermal energy dissipation in systems in which the synthesis of zeaxanthin is very slow. The precise mechanism of the involvement of zeaxanthin in the thermal de-excitation of, presumably, the excited singlet state of chlorophyll is not known. The properties of the zeaxanthin molecule which, in contrast to other carotenoids, allow it to facilitate thermal energy dissipation yet to be determined.

Zeaxanthin-associated thermal energy dissipation appears to occur in all sun-exposed plants, with peak activities during the hours of highest PFD. Sun-exposed leaves possess higher levels of the components of the xanthophyll cycle, and thus a greater capacity for thermal energy dissipation, than do shade leaves. More studies are needed to examine how complete the protection afforded through zeaxanthin-associated energy dissipation is in sun-exposed plants of different species and under different conditions in the field.

ACKNOWLEDGMENT

Much of the recent and unpublished work in our laboratory has been supported by the U.S. Department of Agriculture, Competitive Research Grants Office, Award Number 90-37130-5422.

REFERENCES

1. **Asada, K. and Takahashi, M.,** Production and scavenging of active oxygen in photosynthesis, in *Photoinhibition*, Kyle, D. J., Osmond, C. B., and Arntzen, C. J., Eds., Elsevier, Amsterdam, 1987, 227.
2. **Powles, S. B.,** Photoinhibition of photosynthesis induced by visible light, *Annu. Rev. Plant Physiol.*, 35, 15, 1984.
3. **Demmig-Adams, B. and Adams, W. W., III,** Photoprotection and other responses of plants to high light stress, *Annu. Rev. Plant Physiol. Plant Mol. Biol.*, 43, 599, 1992.

4. **Demmig-Adams, B.,** Carotenoids and photoprotection in plants. A role for the xanthophyll zeaxanthin, *Biochim. Biophys. Acta,* 1020, 1, 1990.
5. **Demmig-Adams, B. and Adams, W. W., III,** The xanthophyll cycle, in *Carotenoids in Photosynthesis,* Young, A. and Britton, G., Eds., Springer-Verlag, Berlin, in press.
6. **Krinsky, N. I.,** Carotenoid protection against oxidation, *Pure Appl. Chem.,* 51, 649, 1979.
7. **Young, A. and Britton, G.,** Carotenoids and stress, in *Stress Responses in Plants: Adaptation and Acclimation Mechanisms,* Alscher, R. G. and Cumming, J. R., Eds., Wiley-Liss, New York, 1990, 87.
8. **Ehleringer, J. R., Björkman, O., and Mooney, H. A.,** Leaf pubescence: effects on absorbance and photosynthesis in a desert shrub, *Science,* 192, 376, 1976.
9. **Koller, D.,** The control of leaf orientation by light, *Photochem. Photobiol.,* 44, 819, 1986.
10. **Britz, S. J.,** Chloroplast and nuclear migration, in *Physiology of Movements, Encyclopedia of Plant Physiology, New Series,* Vol. 7, Haupt, W. and Beinleib, M. E., Eds., Springer-Verlag, Heidelberg, 1979, 170.
11. **Haupt, W. and Scheuerlein, R.,** Chloroplast movement, *Plant Cell Environ.,* 13, 595, 1990.
12. **Hager, A.,** The reversible, light-induced conversions of xanthophylls in the chloroplast, in *Pigments in Plants,* Czygan, F.-C., Ed., Fischer, Stuttgart, 1980, 57.
13. **Siefermann-Harms, D.,** The xanthophyll cycle in higher plants, in *Lipids and Lipid Polymers in Higher Plants,* Tevini, M. and Lichtenthaler, H. K., Eds., Springer-Verlag, Berlin, 1977, 218.
14. **Yamamoto, H. Y.,** Biochemistry of the violaxanthin cycle in higher plants, *Pure Appl. Chem.,* 51, 639, 1979.
15. **Demmig-Adams, B. and Adams, W. W., III,** Carotenoid composition of sun and shade leaves of plants with different life forms, *Plant Cell Environ.,* 15, 411, 1992.
16. **Thayer, S. S. and Björkman, O.,** Leaf xanthophyll content and composition in sun and shade determined by HPLC, *Photosynth. Res.,* 23, 331, 1990.
17. **Jones, B. L. and Porter, J. W.,** Biosynthesis of carotenes in higher plants, *CRC Crit. Rev. Plant Sci.,* 3, 295, 1986.
18. **Duckham, S. C., Linforth, R. S. T., and Taylor, I. B.,** Abscisic-acid-deficient mutants at the *aba* gene locus of *Arabidopsis thaliana* are impaired in the epoxidation of zeaxanthin, *Plant Cell Environ.,* 14, 601, 1991.
19. **Rock, C. D. and Zeevaart, J. A. D.,** The *aba* mutant of *Arabidopsis thaliana* is impaired in epoxy-carotenoid biosynthesis, *Proc. Natl. Acad. Sci. U.S.A.,* 88, 7496, 1991.
20. **Demmig, B., Winter, K., Krüger, A., and Czygan, F.-C.,** Zeaxanthin and the heat dissipation of excess light energy in *Nerium oleander* exposed to a combination of high light and water stress, *Plant Physiol.,* 87, 17, 1988.
21. **Demmig-Adams, B., Winter, K., Winkelmann, E., Krüger, A., and Czygan, F.-C.,** Photosynthetic characteristics and the ratios of chlorophyll, β-carotene, and the components of the xanthophyll cycle upon a sudden increase in growth light regime in several plant species, *Bot. Acta,* 102, 319, 1989.
22. **Siefermann-Harms, D.,** Carotenoids in photosynthesis. I. Location in photosynthetic membranes and light-harvesting function, *Biochim. Biophys. Acta,* 811, 325, 1985.
23. **Costes, C., Burghoffer, C., Joyard, J., Block, M., and Douce, R.,** Occurrence and biosynthesis of violaxanthin in isolated spinach chloroplast envelope, *FEBS Lett.,* 103, 17, 1979.
24. **Siefermann-Harms, D., Joyard, J., and Douce, R.,** Light-induced changes of the carotenoid levels in chloroplast envelopes, *Plant Physiol.,* 61, 530, 1978.
25. **Demmig, B. and Björkman, O.,** Comparison of the effect of excessive light on chlorophyll fluorescence (77K) and photon yield of O_2 evolution in leaves of higher plants, *Planta,* 171, 171, 1987.
26. **Weis, E. and Berry, J. A.,** Quantum efficiency of photosystem II in relation to "energy"-dependent quenching of chlorophyll fluorescence, *Biochim. Biophys. Acta,* 894, 198, 1987.
27. **Krause, G. H.,** Photoinhibition of photosynthesis. An evaluation of damaging and protective mechanisms, *Physiol. Plant.,* 74, 566, 1988.

28. **Demmig-Adams, B., Adams, W. W., III, Winter, K., Meyer, A., Schreiber, U., Pereira, J. S., Krüger, A., Czygan, F.-C., and Lange, O. L.,** Photochemical efficiency of photosystem II, photon yield of O_2 evolution, photosynthetic capacity, and carotenoid composition during the "midday depression" of net CO_2 uptake in *Arbutus unedo* growing in Portugal, *Planta,* 177, 377, 1989.
29. **Björkman, O., Demmig, B., and Andrews, T. J.,** Mangrove photosynthesis: response to high-irradiance stress, *Aust. J. Plant Physiol.,* 15, 43, 1988.
30. **Adams, W. W., III, Smith, S. D., and Osmond, C. B.,** Photoinhibition of the CAM succulent *Opuntia basilaris* growing in Death Valley: evidence from 77K fluorescence and quantum yield, *Oecologia,* 71, 221, 1987.
31. **Adams, W. W., III, Díaz, M., and Winter, K.,** Diurnal changes in photochemical efficiency, the reduction state of Q, radiationless energy dissipation, and non-photochemical fluorescence quenching in cacti exposed to natural sunlight in northern Venezuela, *Oecologia,* 80, 553, 1989.
32. **Krause, G. H. and Weis, E.,** Chlorophyll fluorescence and photosynthesis: the basics, *Annu. Rev. Plant Physiol. Plant Mol. Biol.,* 42, 313, 1991.
33. **Bilger, W. and Björkman, O.,** Role of the xanthophyll cycle in photoprotection elucidated by measurements of light-induced absorbance changes, fluorescence and photosynthesis in leaves of *Hedera canariensis, Photosynth. Res.,* 25, 173, 1990.
34. **Demmig-Adams, B., Adams, W. W., III, Heber, U., Neimanis, S., Winter, K., Krüger, A., Czygan, F.-C., Bilger, W., and Björkman, O.,** Inhibition of zeaxanthin formation and of rapid changes in radiationless energy dissipation by dithiothreitol in spinach leaves and chloroplasts, *Plant Physiol.,* 92, 293, 1990.
35. **Adams, W. W., III, Demmig-Adams, B., and Winter, K.,** Relative contributions of zeaxanthin-related and zeaxanthin-unrelated types of 'high-energy-state' quenching of chlorophyll fluorescence in spinach leaves exposed to various environmental conditions, *Plant Physiol.,* 92, 302, 1990.
36. **Neubauer, C. and Yamamoto, H. Y.,** The Mehler-peroxidase reaction generates the ΔpH that is required for zeaxanthin-related fluorescence quenching, *Plant Physiol. Suppl.,* 96, 119, 1991.
37. **Schreiber, U., Reising, H., and Neubauer, C.,** Contrasting pH-optima of light-driven O_2- and H_2O_2-reduction in spinach chloroplasts as measured *via* chlorophyll fluorescence quenching, *Z. Naturforsch.,* 46c, 635, 1992.
38. **Gilmore, A. M. and Yamamoto, H. Y.,** Zeaxanthin formation and energy-dependent fluorescence quenching in pea chloroplasts under artificially-mediated linear and cyclic electron transport, *Plant Physiol.,* 96, 635, 1991.
39. **Noctor, G., Rees, D., Young, A., and Horton, P.,** The relationship between zeaxanthin, energy-dependent quenching of chlorophyll fluorescence, and trans-thylakoid pH gradient in isolated chloroplasts, *Biochim. Biophys. Acta,* 1057, 320, 1991.
40. **Demmig, B., Winter, K., Krüger, A., and Czygan, F.-C.,** Photoinhibition and zeaxanthin formation in intact leaves. A possible role of the xanthophyll cycle in the dissipation of excess light energy, *Plant Physiol.,* 84, 218, 1987.
41. **Demmig-Adams, B., Winter, K., Krüger, A., and Czygan, F.-C.,** Light response of CO_2 assimilation, dissipation of excess excitation energy, and zeaxanthin content of sun and shade leaves, *Plant Physiol.,* 90, 881, 1989.
42. **Demmig-Adams, B., Winter, K., Krüger, A., and Czygan, F.-C.,** Zeaxanthin and the induction and relaxation kinetics of the dissipation of excess excitation energy in leaves in 2% O_2, 0% CO_2, *Plant Physiol.,* 90, 887, 1989.
43. **Demmig-Adams, B., Winter, K., Krüger, A., and Czygan, F.-C.,** Zeaxanthin synthesis, energy dissipation, and photoprotection of photosytem II at chilling temperatures, *Plant Physiol.,* 90, 894, 1989.
44. **Bilger, W. and Björkman, O.,** Temperature dependence of violaxanthin de-epoxidation and non-photochemical fluorescence quenching in intact leaves of *Gossypium hirsutum* L. and *Malva parviflora* L., *Planta,* 184, 226, 1991.
45. **Yamamoto, H. Y. and Kamite, L.,** The effects of dithiothreitol on violaxanthin de-epoxidation and absorbance changes in the 500-nm region, *Biochim. Biophys. Acta,* 267, 538, 1972.

46. **Bilger, W., Börkman, O., and Thayer, S. S.,** Light-induced spectral absorbance changes in relation to photosynthesis and the epoxidation state of xanthophyll cycle components in cotton leaves, *Plant Physiol.*, 91, 542, 1989.
47. **Demmig-Adams, B. and Adams, W. W., III,** The carotenoid zeaxanthin and 'high-energy state quenching' of chlorophyll fluorescence, *Photosynth. Res.*, 25, 187, 1990.
48. **Demmig-Adams, B., Adams, W. W., III, Czygan, F.-C., Schreiber, U., and Lange, O. L.,** Differences in the capacity for radiationless energy dissipation in green and blue-green algal lichens associated with differences in carotenoid composition, *Planta*, 180, 582, 1990.
49. **Demmig-Adams, B., Adams, W. W., III, Green, T. G. A., Czygan, F.-C., and Lange, O. L.,** Differences in the susceptibility to light stress in two lichens forming a phycosymbiodeme, one partner possessing and one lacking the xanthophyll cycle, *Oecologia*, 84, 451, 1990.
50. **Demmig-Adams, B., Máguas, C., Adams, W. W., III, Meyer, A., Kilian, E., and Lange, O. L.,** Effect of high light on the efficiency of photochemical energy conversion in a variety of lichen species with green and blue-green phycobionts, *Planta*, 180, 400, 1990.
51. **Ruban, A. V., Rees, D., Noctor, G. D., Young, A., and Horton, P.,** Long-wavelength chlorophyll species are associated with amplification of high-energy-state excitation quenching in higher plants, *Biochim. Biophys. Acta*, 1059, 355, 1991.
52. **Adams, W. W., III and Demmig-Adams, B.,** Operation of the xanthophyll cycle in higher plants in response to diurnal changes in incident sunlight, *Planta*, 186, 390, 1992.
53. **Adams, W. W., III, Volk, M., Hoehn, A., and Demmig-Adams, B.,** Leaf orientation and the response of the xanthophyll cycle to incident light, *Oecologia*, 90, 404, 1992.
54. **Khamis, S., Lamaze, T., Lemoine, Y., and Foyer, C.,** Adaptation of the photosynthetic apparatus in maize leaves as a result of nitrogen limitation, *Plant Physiol.*, 94, 1436, 1990.
55. **Morales, F., Abadia, A., and Abadia, J.,** Characterization of the xanthophyll cycle and other photosynthetic pigment changes induced by iron deficiency in sugar beet (*Beta vulgaris* L.), *Plant Physiol.*, 94, 607, 1990.
56. **Schreiber, U., Schliwa, U., and Bilger, W.,** Continuous recording of photochemical and non-photochemical chlorophyll fluorescence quenching with a new type of modulation fluorometer, *Photosynth. Res.*, 10, 51, 1986.
57. **Thayer, S. S., Yamamoto, H. Y., and Björkman, O.,** personal communication.
58. **Bilger, W. and Björkman, O.,** personal communication.
59. **Adams, W. W. III and Demmig-Adams, B.,** unpublished results.

Chapter 5

VITAMIN E, α-TOCOPHEROL

John L. Hess

TABLE OF CONTENTS

0-8493-6328-4/93/$0.00 + $.50

I. INTRODUCTION

The tocopherols, and most specifically α-tocopherol, vitamin E, continues to receive remarkable attention in literature associated with mammalian research. In this chapter, vitamin E refers specifically to **d**-α-tocopherol, MW 430.9. The scientific name, 3,4-dihydro-2,5,7,8-tetramethyl-(4,8,12-trimethyltridecyl)-2-H-1-benzopyran-6-ol, is indeed cumbersome; thus, 5,7,8 trimethyl-tocol has become a generally accepted nomenclature for vitamin E. The related tocol and tocotrienol structures of biological importance are shown in Figure 1.[1,2]

The functions of vitamin E as an antioxidant and as a singlet-oxygen trap are well documented in a variety of symposia and monographs.[3,4] Recently, W. A. Pryor[5] has summarized the literature that addresses the efficacy of vitamin E protection against the pathological effects of ozone. Specific functions for vitamin E have not been established, even though unique responses occur in animals that experience vitamin E deficiency. Furthermore, early reports concerning a potential regulatory function for tocopherols in plants, e.g., initiation of flowering, are of interest,[2,6] but additional experiments to further characterize this or other specific regulatory functions for vitamin E have not been reported.

The occurrence of tocols and tocotrienols has been described for many plants. However, the more important of these compounds appear to be the fully substituted benzoquinone derivatives, α-tocopherol and α-tocotrienol. Vitamin E occurs universally, and its unique role as an antioxidant and stabilizer for biological membranes is related to both its fully substituted benzoquinone ring and fully reduced phytyl side chain.[7,8]

Although the source of dietary vitamin E is from plants, limited experimental data directly establish the antioxidant function of vitamin E in plants.[9] The interest in nutritional requirements for vitamins, and, therefore, commercial sources of vitamin E is a basis for acquiring quantitative data for tocopherol and tocotrienol content in seeds and seed oils.[10-12] The improved stability of plant oils that contain higher quantities of tocopherol has been the basis for selection and breeding of the high tocopherol sunflower, *Helianthus annus* cv Pochnee.[61] A significant effort to understand the biosynthesis of vitamin E in photosynthetic tissue has resulted in a generally accepted pathway of synthesis.[13,14] Since significant quantities of the tocopherols and tocotrienols exist in seed oils, further attention to biosynthesis, accumulation, and function in seeds is warranted.

The role of vitamin E in normal plant development is not understood, and the response of vitamin E to oxidative and other environmental stress is documented with only limited experimental evidence. A perspective of this chapter is to integrate what may be expected to be the function of vitamin E with what is known about its synthesis and occurrence in higher plant tissues.

II. OCCURRENCE

Vitamin E has been found in all photosynthetic organisms,[2,13] and because of its hydrophobic nature, it is always located in membranes of the cell. It generally occurs in highest concentration in green photosynthetic tissue (Table 1), so that its

A.

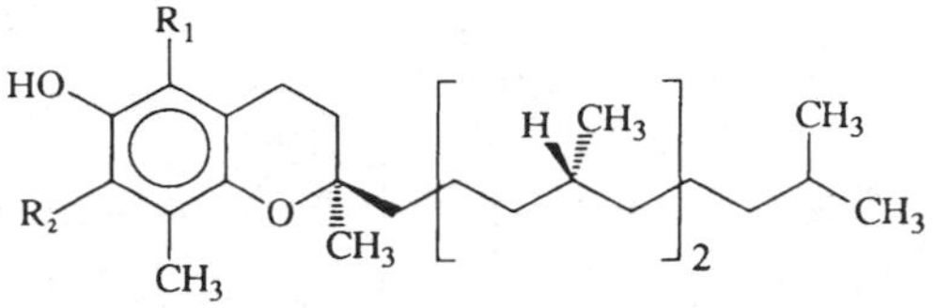

α-tocopherol
$R_1 = CH_3$, $R_2 = CH_3$; 5,7,8-trimethyl tocol

β-tocopherol
$R_1 = CH_3$, $R_2 = H$; 5,8-dimethyl tocol

γ-tocopherol
$R_1 = H$, $R_2 = CH_3$; 7,8-dimethyl tocol

δ-tocopherol
$R_1 = H$, $R_2 = H$; 8-methyl tocol

B.

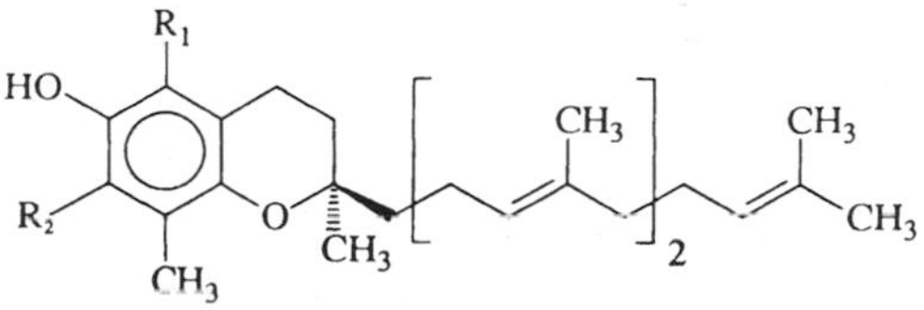

α-tocotrienol
$R_1 = CH_3$, $R_2 = CH_3$; 5,7,8-trimethyl tocotrienol

β-tocotrienol
$R_1 = CH_3$, $R_2 = H$; 5,8-dimethyl tocotrienol

γ-tocotrienol
$R_1 = H$, $R_2 = CH_3$; 7,8-dimethyl tocotrienol

δ-tocotrienol
$R_1 = H$, $R_2 = H$; 8-methyl tocotrienol

FIGURE 1. The family of the tocopherols (A) and the tocotrienols (B) that are differentiated by the unsaturation in the diterpenoid side chain. Nomenclature is based on methylation pattern.[1,2]

presence in the chloroplast membrane is expected. However, there are no quantitative data that establish the actual distribution of vitamin E among the plant cell membranes. In a brief report, Yerin et al.[15] concluded that the bulk of the vitamin E partitions into the lipid phase of the chloroplast membrane. This result is consistent with the studies on the location of vitamin E in phospholipid vesicles. The phytyl side chain is embedded within the bilayer, and the benzoquinone ring, which interacts with the carbonyl of the glycerol esters, establishes the orientation of the vitamin E molecule (Figure 2).[8]

TABLE 1
Concentrations of the Major Tocopherols in Various Tissues[a]

A. Data Reported on the Basis of Tissue Wet (Fresh) Weight μg per gfw

Source	α-Tocopherol	γ-Tocopherol	Ref.
Abutilon sp. variegated			
Green	43.5	12.3	2
Yellow	9.1	7.4	
Avena sativa			
14-day seedling	17	—	58
+ 4 days drought	25	—	
Coleus blumei	32.3	14.8	2
Coleus hybridus schwartzei variegated			
Green	19.8	9.4	2
Yellow	6.9	5.0	
Deschampsia flexuosa			
14-day seedling	17	—	58
+ 4 days drought	33	—	
Ficus elastica variegated			
Green	34.7	5.2	2
Yellow	11.8	9.5	
Hevea brasiliensis	71.2	18.4	2
Holcus lanatus			
14-day seedling	5.9	—	58
+ 4 days drought	14	—	
Hordeum vulgare			
Mature leaf	15.7	8.3	2
14-day seedling	26	—	58
+ 4 days drought	62	—	
Ilex aquifolium variegated			
Green	169	43	2
Yellow	7.8	4.6	
Ligustrum vulgare	41.4	10.3	2
Melia azedarach			
Green	204	—	25
Senescing	510	—	
Peperomia americana variegated			
Green	30.1	5.1	2
Yellow	2.4	2.9	
Phaseolus areus	39.7	29.4	2
Picea abies			
Primary needles	43	—	18
Roots	2.5	—	
Etiolated needles	9	—	
Current year	8.5	—	
1 year	100	—	
2 year	150	—	
3 year	215	—	
Seed	60	—	
Cotyledon	98	23	

TABLE 1 (continued)
Concentrations of the Major Tocopherols in Various Tissues[a]

A. Data Reported on the Basis of Tissue Wet (Fresh) Weight μg per gfw

Source	α-Tocopherol	γ-Tocopherol	Ref.
Picea bicolor			
Current year	3.8	—	18
1 year	99	—	
2 year	93	—	
Picea koyamai			
Current year	9.3	—	18
1 year	107	—	
2 year	153	—	
Poa trivialis			
14-day Seedling	3.9	—	58
+ 4 days drought	6.7	—	
Polygonum cuspidatum	28.4	12.2	2
Solanum tuberosum			
Fresh	0.10	—	16
Stored 40 weeks, 3°C	0.29	—	
Rumex sanguineus	16.9	6.9	2
Scenedesmus obliquus	41	—	26
Mutant PS28	none detected	—	
Sinapis alba	250	—	53
Spinacea oleracea	24.3	4.2	2
Taxus baccata	180	—	53
Trifolium ripens	160	—	53
Triticum aestivum			
14-day seedling	8.6	—	58
+ 4 days drought	35	—	
Tussilago farfara	12.1	2.1	2
Vinca major variegated			
Green	22	—	25
Senescing	340	—	
Zea mays	2.1	2.1	2

B. Data Reported on the Basis of Tissue Dry Weight μg per gdw

Source	α-Tocopherol	Other tocopherol or tocotrienols	Ref.
Abies alba			
1 year needle	200	—	55
4 year needle	700	—	
Abutilon theophrasti			
Cotyledon	500	—	60
Amaranthus retroflexus			
Cotyledon	100	—	60
Cassia obtusifolia			
Cotyledon	600	—	60

TABLE 1 (continued)
Concentrations of the Major Tocopherols in Various Tissues[a]

B. Data Reported on the Basis of Tissue Dry Weight μg per gdw

Source	α-Tocopherol	γ-Tocopherol	Ref.
Carthamus tinctorius			
Callus culture	148	4.4 β-tocopherol	19
Supplemented with phytol	287	30 β-tocopherol	
Chenopodium album			
Cotyledon	120	—	60
Datura stramonium			
Cotyledon	830	—	60
Elaeis guineensis			
Leaflet	3200–5600	—	17
Fagopyrum esculentum			
Cotyledon	280	—	60
Fagus silvatica			
1 month leaf	300	—	55
5 month leaf	2300	—	
Hevea brasiliensis (latex)			
Tocotrienol	Trace	120 α-tocotrienol 250 γ-tocotrienol	27
Hordeum sativum			
Seed 23% moisture	13	41 α-tocotrienol	11
Stored 11 months	5.0	6.0 α-tocotrienol	
Stored 11 months + CO_2	9.0	33 α-tocotrienol	
Helianthus annum			
Seed	860–950	—	12
Hordeum sativum			
Seed	9.4	19 tocotrienol	10
74 day plant	67	—	
Green ear	78	5.7 tocotrienol	
Ripe ear	10	17 tocotrienol	
Ipomoea purpurea			
Cotyledon	100	—	60
Kalanchoe crenata	625	—	30
Dark-grown callus	20	—	
Green callus	72	—	
Lepidium sativum	125	—	53
Medicago sativa	145	—	53
Cotyledon	100	—	60
Pisum sativum			
Seed	6.5	158 γ-tocopherol	10
31 day plants	127	—	
New pods	456	—	
New seed	Trace	23 γ-tocopherol	
8 day etiolated plant	51	86 γ-tocopherol	
Sinapis alba	250		53
Cotyledon	500	—	60
		—	

TABLE 1 (continued)
Concentrations of the Major Tocopherols in Various Tissues[a]

B. Data Reported on the Basis of Tissue Dry Weight μg per gdw

Source	α-Tocopherol	γ-Tocopherol	Ref.
Solanum tuberosum			
Fresh	0.10	—	16
Stored 40 weeks, 3°C	0.29	—	
Spinacea oleracea		—	
Fresh	170	—	53
Wilted	256	—	53
Taxus baccata	180	—	53
Trifolium ripens	160	—	53
Triticum vulgare			
Seed	14	19 Tocotrienol	10
38 day plant	82	—	
Ripe ear	5.7	8.2 Tocotrienol	
14 day etiolated plant	57	—	
Zea mays			
Seed	5.6	95 γ-tocopherol	10
87 day plant	54	36 γ-tocopherol	
Unripe cob (136 day)	73	46 γ-tocopherol	
16 day etiolated plant	70	99 γ-tocopherol	

[a] The summary of quantitative data is representative of that reported using paper, thin layer (TLC), and high performance liquid chromatography (HPLC). The hyphen indicates that no data were reported and, unless otherwise specified, data are from mature leaf tissue. Generally no statistical parameters occur in the earlier literature.

The relative content of vitamin E varies from low concentrations in potato tubers (*Solanum tuberosum* L.) (150 to 300 ng per gram fresh weight [gfw]) to high concentrations in oil palm (*Elais guineensis*) leaflets (3 to 6 mg per gfw).[16,17] The occurrence of vitamin E in roots, seeds, and etiolated tissues, as well as in green leaf tissue, indicates that it can be synthesized in the nonchlorophyll-containing tissues.[2,10,18] In variegated leaf tissue, vitamin E content ranges from 170 μg per gfw (green) and 8 μg per gfw (yellow) in *Ilex aquifolium* to 20 μg per gfw (green) and 7 μg per gfw (yellow) in *Coleus hybridus*.[2] In these yellow tissues, there occur significantly greater ratios of γ-tocopherol to α-tocopherol. The synthesis of vitamin E in nonphotosynthetic safflower (*Carthamus tinctorius*) callus culture also occurs,[19] and provides a model in which to explore factors that influence or regulate tocopherol synthesis.

III. BIOSYNTHESIS

Vitamin E belongs to a family of antioxidants that includes four methylated tocols, substituted with a phytyl chain, and the analogous tocotrienols, substituted with a geranylgeranyl chain (Figure 1).[2,14] Of the tocopherols, vitamin E, is the major constituent and contains the fully substituted benzoquinone ring, and is

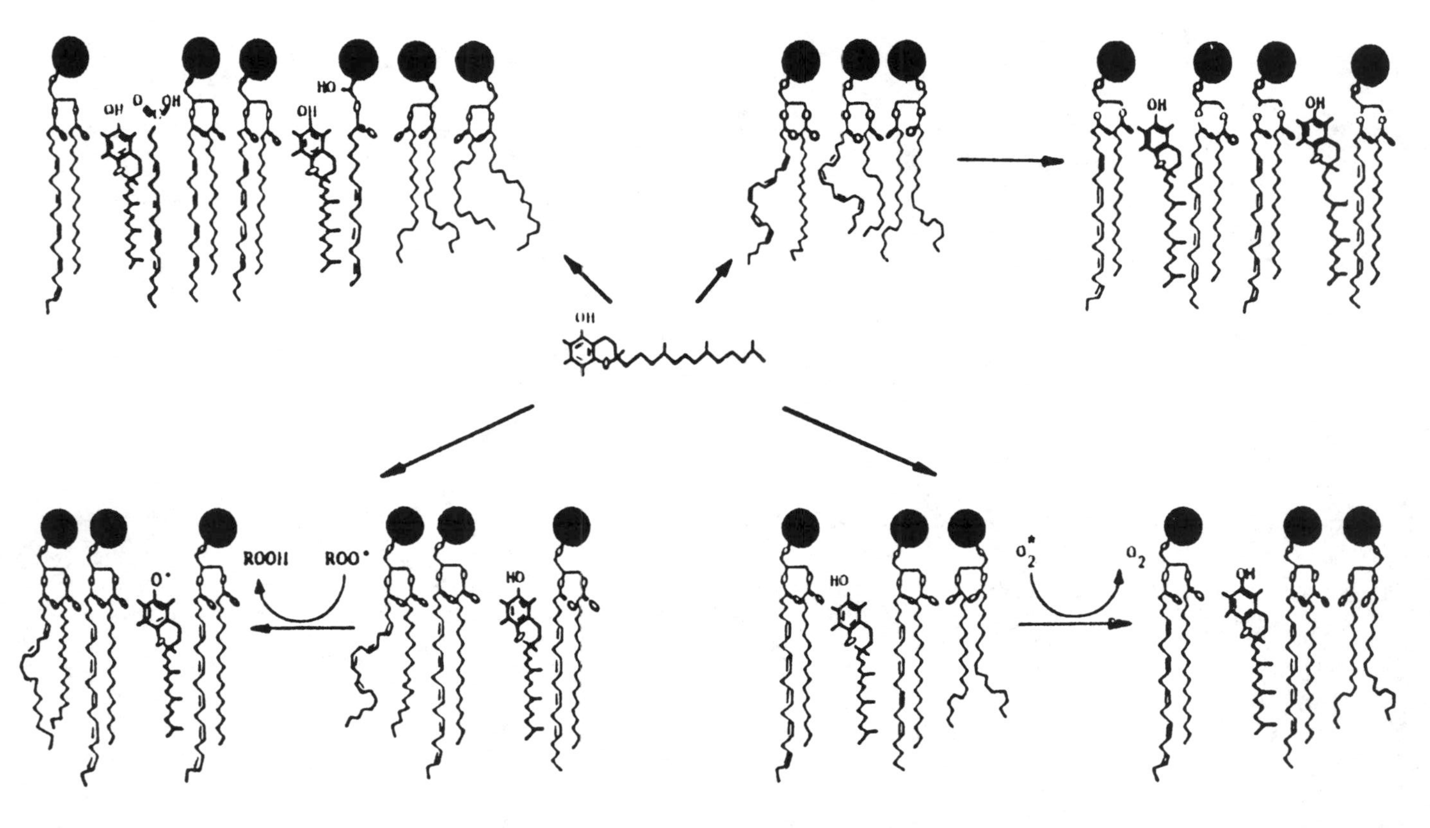

FIGURE 2. Models for the association of vitamin E with phospholipid bilayers and possible mechanisms for stabilizing these membranes' models.[33]

identified as 5,7,8-trimethyl tocol. The β-tocopherol (5,8-dimethyl tocol) and γ-tocopherol (7,8-dimethyl tocol) may exist as intermediates in the synthesis of vitamin E from the common precursor, δ-tocopherol (8-methyl tocol) (Figure 3). Although one might interconnect the pathways for the biosynthesis of these two classes of antioxidants,[20] it is likely that there are similar but distinct methylation and cyclization steps for the synthesis of the tocols and the tocotrienols. Their synthesis may depend on both the site of synthesis and the availability of either phytyl pyrophosphate or geranylgeranyl pyrophosphate. The committed step for tocopherol biosynthesis is catalyzed by homogentisate decarboxylase phytyltransferase.[21] It is this cyclization step that forms the physiologically active **d**-isomer of the tocochromanols.

In higher plants, α-tocopherol is synthesized in chloroplasts and proplastids, as proposed in Figure 3.[13,14] The aromatic ring formed in the chloroplast is derived from the shikimic acid pathway intermediate, homogentisic acid,[22] which reacts with phytylpyrophosphate to yield 2-methyl-6-phytyl-1,4-benzoquinol, CO_2, and pyrophosphate. Phytyl pyrophosphate formation from geranylgeranyl pyrophosphate depends on NADPH and a chloroplast envelope enzyme.[23] Homogentisic acid biosynthesis occurs in the chloroplast[13,21] and the plastids of nonphotosynthetic tissue.[28] That available phytol/phytylpyrophosphate serves as a limiting precursor for α-tocopherol synthesis is supported by the increased tocopherol content in senescing leaves. As chlorophyll degradation occurs and releases phytol to the terpenoid pool, vitamin E synthesis increases.[25] Furthermore, the mutant of *Scenedesmus obliquus,* PS28, was unable to form α-tocopherol, but it did form chlorophyll containing a geranylgeranol ester.[26] Hence, it was concluded that the reduced diterpenoid was required for tocopherol synthesis. The latex of *Havea brasiliensis* and some seeds contain significant quantities of the tocotrienols and only minor amounts of vitamin E.[10,27] This distribution in nonphotosynthetic tissue is consistent with access to a source of only geranylgeranyl pyrophosphate for tocotrienol synthesis. In these tissues, plastid reduction of the diterpene might not occur.

Remaining steps in the biosynthesis generally depend on the exclusive use of the 6-phytylbenzoquinol. The tocotrienols presumably would use the 6-geranylgeranylbenzoquinol. Low quantities of the 5- and 3-phytylbenzoquinone isomers may occur as products from the nonspecific side reactions.[14]

Two, *S*-adenosylmethionine-dependent methylation reactions and a cyclization reaction are required to complete α-tocopherol biosynthesis. In spinach (*Spinacia oleracea* L.), it has been established that the preferred pathway is (1) C-methylation in the C2 position; (2) cyclization to form the **d**-7,8-dimethyl tocol, γ-tocopherol; and (3) a second methylation, at position C5 of the tocol ring, leads to α-tocopherol as the final product. The occurrence of δ- and β-tocopherols in some species[10,20] implicates the alternate route in which cyclization precedes either methylation step. The methyltransferase(s) appear to be specific with regard to the position that is methylated, but are much less fastidious about the quinol, tocol, or tocotrienol form of the substrate. A γ-tocopherol methyltransferase has been purified to homogeneity from *Capsicum annum* fruit chromoplasts.[28]

Janiszowska reported that chloroplasts are the main site of tocopherol synthesis, but that if phytol is available, then the microsomes are also a possible site of

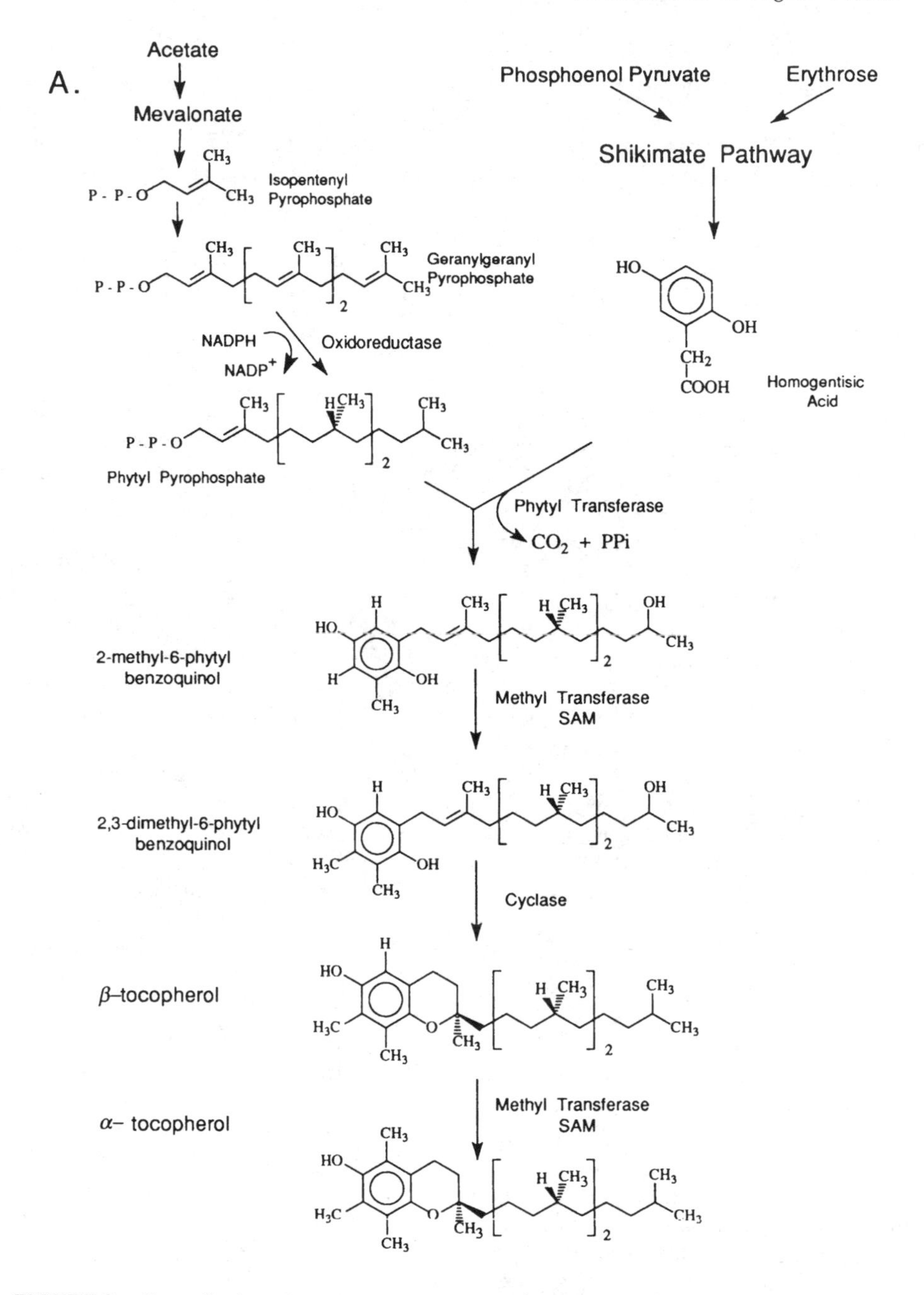

FIGURE 3. Generalized, preferred biosynthetic pathway of tocopherol synthesis in chloroplasts (A)[14] and possible pathways for tocotrienol (B) synthesis in higher plants.[22] No evidence supports the occurrence of the cyclization of the benzoquinol to the chromanol ring outside of the intact chloroplast. Methyltransferases occur both within the chloroplast and with the endoplasmic reticulum. *S*-adenosylmethionine (SAM) serves as the methyl donor in these transferase-catalyzed reactions.

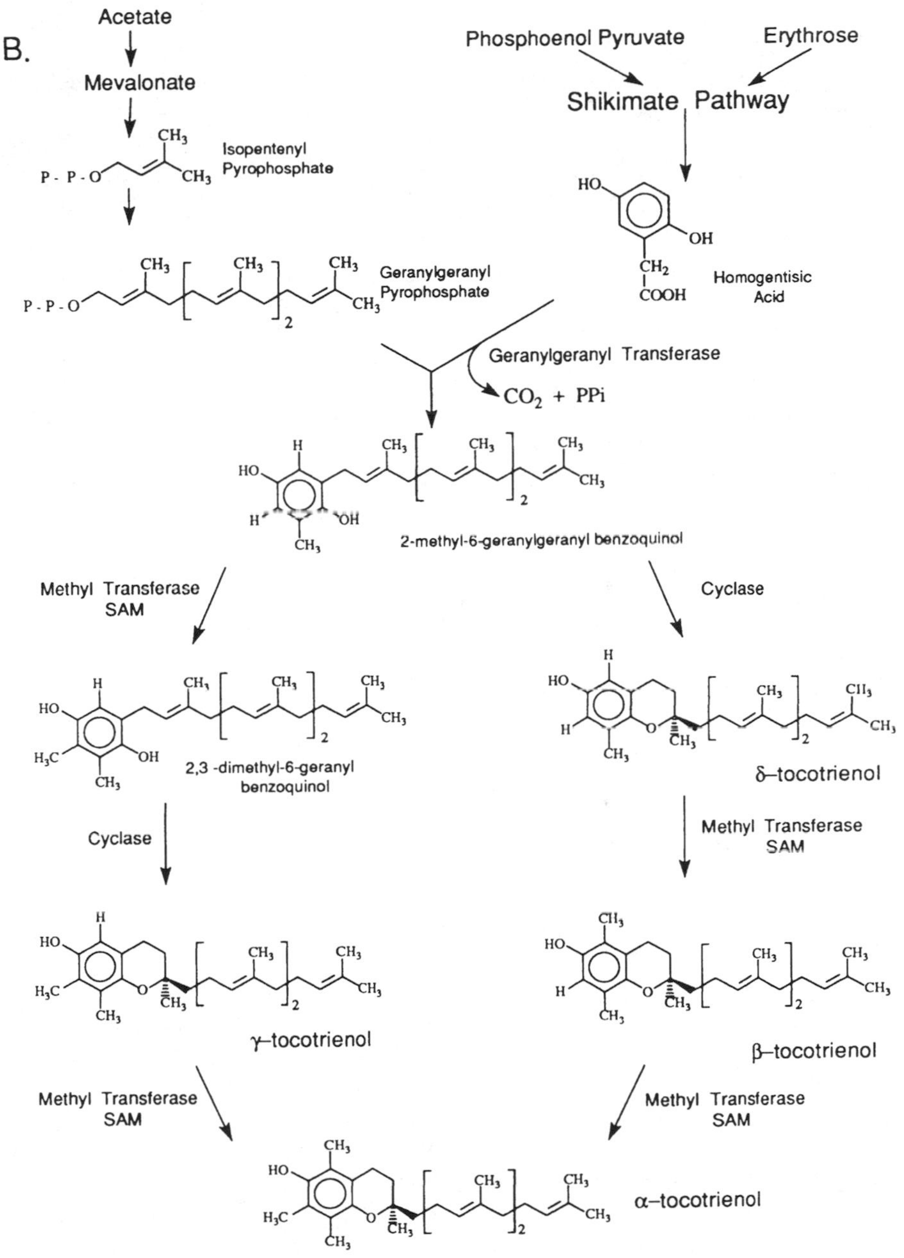

B.

Acetate

Mevalonate

Isopentenyl Pyrophosphate

Geranylgeranyl Pyrophosphate

Phosphoenol Pyruvate

Erythrose

Shikimate Pathway

Homogentisic Acid

Geranylgeranyl Transferase

CO_2 + PPi

2-methyl-6-geranylgeranyl benzoquinol

Methyl Transferase SAM

Cyclase

2,3 -dimethyl-6-geranyl benzoquinol

δ–tocotrienol

Cyclase

Methyl Transferase SAM

γ–tocotrienol

β–tocotrienol

Methyl Transferase SAM

Methyl Transferase SAM

α–tocotrienol

FIGURE 3B.

synthesis.[29] Most likely, tocopherol is not synthesized by the mitochondria. Its occurrence in the mitochondria is likely derived by uptake from the microsomal membrane.[29] Thomas and Stobart[30] reported that vitamin E synthesis correlated with greening in callus cultures of *Kalanchoe crenata*. They also recognized that incorporation of ^{14}C-mevalonate into vitamin E occurred with a greater specific activity in the membrane fraction than in intact chloroplast. These data suggest that sites outside of the chloroplast may be competent with regard to vitamin E synthesis, although the composition of this membrane fraction (3000 to 20,000 *g*) was not further characterized, and may have contained chloroplast membrane fragments, as well as endoplasmic reticulum. Heintze et al.[31] have further demonstrated that, although the source of carbon may differ, both immature plastids and mature chloroplasts are competent with regard to terpenoid biosynthesis. The methyltransferase and the geranylgeranyl reductase are envelope enzymes, and the cyclase is presumably a stromal enzyme, since cyclization has been observed only with intact chloroplasts. No experiments reveal the occurrence of an extraplastidic cyclase, although this enzyme must be required for the synthesis of the tocotrienols (Figure 3B). These data are consistent with the ability of exogenous phytol to bring about increased α-tocopherol synthesis in cell suspension cultures derived from callus tissue grown from immature safflower petals.[19] In summary, it appears that the availability of homogentisate, *S*-adenosylmethionine, and phytol or geranylgeranol supplies necessary carbon for the synthesis of the tocopherols and tocotrienols.

Hence, vitamin E biosynthesis may be regulated by manipulating either the shikimic acid pathway and/or the terpenoid pathway. It is likely that tocotrienol synthesis occurs mainly outside of the chloroplast, and that vitamin E synthesis is restricted to the chloroplast or plastid due to the specific requirements for the reduction of geranylgeranyl pyrophosphate. However, the question originally posed by Threlfall: "Are the tocopherol synthesizing sites outside as well as inside the chloroplast or are all the tocopherols synthesized inside and then transported out?"[32] remains unanswered.

IV. FUNCTIONS

A. MEMBRANE STABILIZATION

From measurements of UV spectra and ^{1}H nuclear magnetic resonance (NMR) measurements, Kagen and colleagues[33,34] have established that complexes form between vitamin E and free fatty acids. They established that hydrogen bonding stabilizes the association between the chromanol ring and the carboxyl group of the fatty acid. This complex may also be stabilized by hydrophobic interactions between the fatty acid acyl chain and the chromanol ring. There are no experiments reported that indicate how the complex with fatty acids might affect the antioxidant potential of the vitamin E.

Vitamin E partitions into phospholipid vesicles, and is oriented, as noted in Figure 2, within the bilayer so that intimate associations of the phytyl chain with the inner regions stabilize the association.[8] A very high, lateral diffusion rate for α-tocopherol in phospholipid bilayers is consistent with a dynamic rather than a static environment for the vitamin.[8,35] In lipid bilayers, unsaturated fatty acids increased the stability of the complex through interactions with the phytyl chain of vitamin E.[33] In these experiments, it is the ability of the tocopherol to influence lipid organization that stabilizes the bilayer, and not its antioxidant potential.[8] This same property can also minimize disordering effects of phospholipid hydrolysis products, particularly fatty acids, that may occur during membrane turnover or in response to external stress, e.g., increased lipid hydrolysis.

Finally, the role of the phytyl chain in stabilizing membrane structures that contain unsaturated fatty acids may explain the greater antioxidant properties of the tocopherols compared to the tocotrienols. The tocotrienol, with its unsaturated side chain, provides less stability to the bilayers containing polyunsaturated fatty acids than does vitamin E. Baszynski[36] completed a series of experiments in which vitamin E uniquely restored PSI electron transport activity in heptane-extracted chloroplasts. Although addition of phytol less effectively restored activity, it was apparent that antioxidant function was not the only purpose served by vitamin E. These data confirm that vitamin E function depends on structural properties that affect membrane function.

B. ANTIOXIDANT ACTIVITY

Generally, vitamin E is considered to be an effective quenching agent for both singlet O_2 and for alkyl peroxides.[6] Plants, particularly photosynthetic tissues, contain significant quantities of β-carotene and carotenoids. Mascio et al.[37] demonstrated the superior ability of the carotenoids to quench singlet oxygen compared to vitamin E. Van Hasselt et al.[38] also reported that both vitamin E and β-carotene effectively protected chlorophyll, in acetone solutions, against red-light-induced degradation. However, on a molar basis, vitamin E was less effective than was β-carotene, and, although their effects were additive at low concentrations, the protection in the presence of both was never as effective as observed with either compound alone. Also, phosphatidylcholine increased the effectiveness of both vitamin E and β-carotene in protecting chlorophyll *a* from photodestruction. This effect may be caused by improved structural associations that the lipid provided. In higher plants, then, vitamin E may not function as the preferred singlet-oxygen trap.

Using a series of tocopherol analogs, Skinner and Parkhurst[39] established the efficacy of vitamin E to function as an antioxidant *in vitro* (protection of β-carotene degradation in corn-oil solutions at 50°C) and *in vivo* (prevention of necrotic liver degeneration). These experiments established requirements for the fully substituted ring and for the presence of the phytyl side chain for maximum, effective protection *in vivo*.

Burton and Ingold,[7] in summarizing much research on vitamin E, identified the unique capacity for vitamin E to function as an effective antioxidant by interrupting carbon-centered reactions. The normal processes of free radical-dependent autooxidation reactions include initiation, propagation, and termination, and have been characterized in solutions. Experiments *in vitro* established that the structure of vitamin E approximates optimal properties for trapping peroxyl radicals. The stabilization of the tocopheroxyl radical formed in the rate-limiting reactions between vitamin E and a peroxy radical is insured by the properties of the fully substituted benzoquinone ring. This stability is consistent with the rapid reaction with alkyl-peroxyl radicals (2.4×10^6 M^{-1} s^{-1}).[40] This reactivity effectively interrupts the propagation of a free-radical chain reaction, and makes vitamin E a very effective free-radical trap. For comparison, the oxidation of vitamin E by superoxide anion occurs at a rate of only 6×10^3 M^{-1} s^{-1}.[41] This latter rate compares very unfavorably with the superoxide dismutase catalyzed rate for O_2^- reduction (2×10^8 M^{-1} s^{-1}). Hence, vitamin E is well suited for its antioxidant function for radicals that partition into the lipid phase, and will react slowly with free radicals as an antioxidant in the aqueous phase.

The phytyl-chain interaction with fatty-acid components of biological membranes and the association of the chromanol ring with carbonyl groups of the lipid esters orients vitamin E. The rapid, lateral diffusion rate for vitamin E in vesicles would, along with rapid movements of peroxyl radicals formed within the hydrophobic core of the membrane as they move to the exterior, facilitate reactions of the radicals with the reactive oxygen of the chromanol ring.[8] This ring oxygen is near the surface of the bilayer, but not fully exposed to the aqueous environment.

That vitamin E functions near the surface of membranes is established through the synergy observed when both vitamin C and E are used to inhibit peroxidations in synthetic phospholipid vesicles. If radicals are generated in the aqueous phase, then vitamin C is itself a good antioxidant. It is, however, a poor antioxidant if the radicals are generated within the lipid phase.[42,43] In these studies, the lipid-soluble diazo compound, 2,2′-azobis(2,4-dimethylvaleronitrile) (AMVN), is used to generate free radicals in the lipid phase. In water/alcohol/acetone solutions, vitamin C has the capacity to reduce the α-tocopheroxyl radical back to the alcohol with a rate constant of 1.55×10^6 M^{-1} s^{-1}.[44] Hence, the greater antioxidant activity in these mixtures of vitamin E and C can be expected from the greater turnover of the tocopheroxyl radical and the alcohol. The ability for vitamin C to regenerate vitamin E directly in the presence of alkylperoxy radicals was elegantly demonstrated by Niki et al.[45] in alcohol solutions of methyl linoleate, vitamin E, and vitamin C. Reactions were initiated with the azonitrile, AMVN, and the content of both vitamins was quantified. The overall capacity to inhibit the linoleate peroxidation depended on the total concentrations of vitamin E and C. In the presence of both vitamins, the rate of α-tocopheroxyl radical formation determined the rate by which linolenate peroxidation was limited. It was further demonstrated that vitamin E was maintained until the vitamin C had been completely oxidized; then vitamin E depletion occurred.[45]

Although two molecules of monodehydroascorbate dismutate to form ascorbate and dehydroascorbate, the occurrence of monodehydroascorbate reductase in plant chloroplasts provides a potentially important catalytic link between vitamin E, vitamin C, and available reducing equivalents in the cell (NADPH or NADH).[46] This enzyme usually occurs in low concentration in plant extracts,[46-48] and is typically considered to function in the H_2O_2 scavenging system of the plant. The enzyme occurs in the chloroplast stroma and contains a flavin, FAD. It differs from the *Neurospora* enzyme, which contains no prosthetic group, and is specific for NADH. In mammalian cells, a similar enzyme occurs in microsomal and mitochondrial membrane preparations that reacts with NADH, contains FAD, but is not soluble

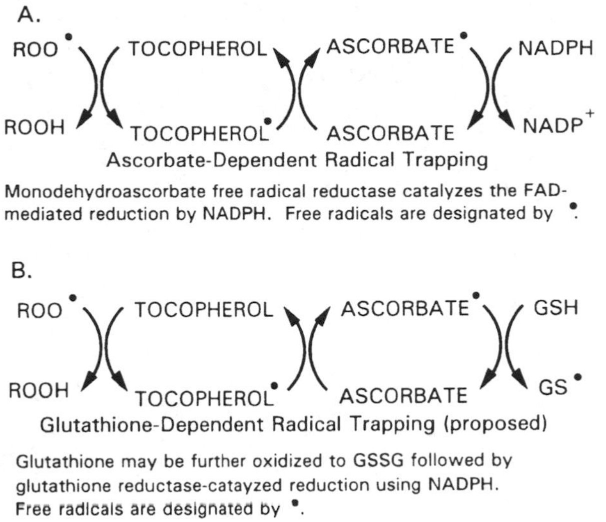

A.

ROO• TOCOPHEROL ASCORBATE• NADPH

ROOH TOCOPHEROL• ASCORBATE NADP+

Ascorbate-Dependent Radical Trapping

Monodehydroascorbate free radical reductase catalyzes the FAD-mediated reduction by NADPH. Free radicals are designated by •.

B.

ROO• TOCOPHEROL ASCORBATE• GSH

ROOH TOCOPHEROL• ASCORBATE GS•

Glutathione-Dependent Radical Trapping (proposed)

Glutathione may be further oxidized to GSSG followed by glutathione reductase-catayzed reduction using NADPH. Free radicals are designated by •.

FIGURE 4. Potential reactions for trapping alkylperoxyl free radicals by vitamin E (tocopherol) and the reduction of vitamin E back to the free quinol. (A) The coupling to vitamin C (ascorbate) is well documented using the monodehydroascorbate reductase.[46] (B) The potential for GSH to reduce monodehydroascorbate has been demonstrated in mammalian liver,[50] but has not been reported in plants.

as is the stromal enzyme of the chloroplast. The ping-pong mechanism allows for a two-electron transfer from NAD(P)H to FAD, and for sequential reduction of two molecules of monodehydroascorbate radicals.[49] Since the reaction between the tocopheroxyl radical and vitamin C will result in a one-electron transfer from the ascorbate, the monodehydroascorbate product will be an effective substrate for this enzyme. This catalysis (Figure 4A) should improve the ability of the chloroplast to regenerate vitamin E at the expense of NAD(P)H, and at the same time, minimize the pool sizes of either the tocopheroxyl or monodehydro-ascorbate radicals.

McCay et al.[50] have recently identified a heat-sensitive factor that may function with glutathione (GSH) to reduce the tocopheroxyl radical in microsomal preparations. This observation is consistent with the ability of glutathione to reduce the vitamin E radical in alcohol solutions.[51] Recently, Goin et al.[52] established the occurrence of thiyl radicals in reactions between GSH and 1,4-naphthoquinones that could account for the mechanism by which GSH could react with the tocopheroxyl radical. Although compartmentation and concentrations of GSH favor the reaction (Figure 4B), there is no evidence for a direct participation of GSH in reducing the vitamin E radical in plants. This possibility should stimulate further interest in characterizing thiyl radicals in plants.

V. RESPONSES TO ENVIRONMENTAL FACTORS

Tocopherol exists as a unique antioxidant in the membranes of animal cells. As discussed elsewhere in this volume, plant membranes contain other potential scavengers for singlet oxygen and free radicals. Hence, exploring and describing a unique function for vitamin E is difficult. However, increased research effort is currently being directed to understanding how vitamin E functions as a component of the antioxidant system that responds to environmental factors.

A. NORMAL DEVELOPMENT/LIFE CYCLE

The early work of Green[10] quantified the occurrence of tocopherols in tissues at various stages of the life cycle in plants, including the seed, mature and immature leaf, fruit, and comparisons of dark- and light-grown leaf tissue. Although Green's techniques were not based on the current high performance liquid chromatography (HPLC) procedures, quantities that were reported agree well with those reported in the current literature (Table 1). Significantly, dark-grown tissues of wheat (*Triticum vulgare*), barley (*Hordeum sativum*), and corn (*Zea mays*) all accumulated tocopherols, and the distribution in wheat plants of α-tocopherol varied from a high of 95 to 100% in mature, field-grown leaf tissue in wheat to 40 to 50% in germinating shoots in the dark, with the remaining percent existing as one of the tocotrienol isomers. Total amounts for the tocopherol pool was greatest in the mature leaf, 90 μg per gdw and lowest in the dark-grown tissue, 40 to 50 μg per gfw. In pea (*Pisum sativum*) plants, the content was almost 100% α-tocopherol in various mature photosynthetic tissues, leaves, and pods. The immature plants, whether light or dark grown, had similar levels of total tocopherols, although in the dark-grown plants, 70 to 80% of the pool was γ-tocopherol. The pea seed also contained >90% γ-tocopherol. In maize plants, on the other hand, only mature leaves had α-tocopherol content at 60% of the total. As with pea tissues, γ-tocopherol was the other major constituent.

Recognizing the limited specific information about vitamin E in woody plants, Franzen et al.[18] described the distribution of Vitamin E in Norway spruce (*Picae abies*) seedlings. Results for the initial 80 days following germination revealed that vitamin E content generally correlated with biomass production, and was the main tocopherol in all organs. Root, hypocotyl, cotyledons, and primary needles were evaluated. Typically, roots had the lowest quantity of vitamin E, and the highest amounts were in the photosynthetic tissue. Vitamin E was synthesized in etiolated seedlings needles, as well. Small quantities of γ- and β-tocopherol, as well as α-tocotrienol, were also detected. Collectively, these data establish that tocopherol content increased with age of leaf tissue, that tocopherol synthesis occurred in both the light- and dark-grown tissue, and that the relative amount of α-tocopherol varied among species, as does the relative contribution of the tocotrienols to the total chromanol pool. No data exist about how enzyme activities might affect pool sizes of vitamin E.

Booth[53] had observed that vitamin E content generally increased in leaves detached from the plant and held in the dark or allowed to wilt in the light. In these experiments, the content in fresh *Spinicea oleracea* was 170 μg per gdw, and increased after 24 hr of wilting to 256 μg per gdw. Data regarding changes in dry weight per leaf area were not provided, and no significant changes in chlorophyll content were observed. Rise et al.[25] more recently reported the increase of vitamin E in senescing leaves of various species. Leaves included field-grown samples and plants grown in controlled environments. Their primary interest was to quantify chlorophyll degradation in senescing tissues. They discovered in *Vinca minor* and *Melia azedarach* (China tree), however, that as chlorophyll content decreased, the quantity of vitamin E increased. The total phytol pool remained nearly constant until very near the death of the leaf. They also reported an increased vitamin E content in citrus (*Citrus sinensis* L.) tissues, both in orange leaves and in green fruit exposed to ethylene. These data may be of general importance in understanding aspects of chlorophyll degradation in plants. This reciprocal relationship between chlorophyll and vitamin E content, however, was not detected in parsley (*Petroselinum sativum* L.), celery (*Apium graveolens* L.), or tobacco (*Nicotiana tobacum* L.). The inability to detect vitamin E in some species may require further development of specific extraction protocols that could include the addition of hydrophobic antioxidants like butylated hydroxytoluene (BHT) through lipid extraction and separation steps.[54]

Kunert and Ederer[55] have quantified the relationship between vitamin E and vitamin C in aging leaf tissue for annual beech (*Fagus silvatica* Mill.) and fir (*Abies alba* Mill.). The increase in vitamin C to a maximum concentration of 3 mg per gdw occurs by the second month of age in the beech leaf, while vitamin E does not reach a maximum concentration of 2 mg per gdw until the fourth month of age. The vitamin C to E ratio is at a minimum value of one during the last 2 months of leaf age in this annual cycle for the beech leaf. For the conifer leaf, vitamin C increases to a maximum value during the second year (2.6 mg per gdw), while vitamin E tends to continue to increase over the first 3 years of the 5-year life cycle of these leaves. For the fir leaf, the vitamin C to E ratio was at a minimum value of four during the last year of the life cycle of the leaf. In these experiments, malondialdehyde formation was also quantified, and revealed that maximum lipid oxidation occurred during the period of the leaf's life cycle, which corresponded to a minimum vitamin C to E ratio. What remains unexplained is whether the increased lipid peroxidation causes the increase in vitamin E in the presence of relatively constant vitamin C levels, or if some other signal is required. It seems unlikely that the "potency of the antioxidant system" declined as suggested by the authors, since the total vitamin content was greatest at the time of leaf senescence. Since only single time points are reported for each year of the fir leaf, no information is available regarding seasonal variation of vitamin content.

B. CHILLING STRESS/HARDENING

A variety of factors are known to contribute to cold acclimation in plants. These include changes in protein composition; lipids — particularly increased unsaturated fatty acids as they affect membrane fluidity; and accumulation of various solutes, sugars, and amino acids.[56] A successful response to decreased temperature (chilling) in leaves may be attributed to the capacity of the leaf to dissipate excess excitation energy. It is expected that, at lower temperatures, the environment becomes more oxidizing. Hence, less damage and greater chilling tolerance is expected if the oxygen concentration is decreased or if the antioxidant metabolism is increased.

In both cucumber (*Cucumis sativus* L. cv Ashley) and pea (*Pisum sativum* L. cv Early Alaska), Wise and Naylor[57] explored the response of pigments and the vitamin E pool in excised leaves maintained at 5°C. Plants were grown at 26/20°C. They reported a light-dependent decrease in vitamin E from 50 nmol per gfw to 10 nmol per gfw after a 3-hr treatment only in the temperature-sensitive cucumber leaf. The level in pea remained at 65 nmol per gfw throughout the treatment. Losses were also reported for the carotenoids and chlorophyll in the cucumber leaf. They reported that ethane production, an indicator of hydrocarbon oxidation, was 50% less in the chilled leaves maintained in 100% N_2. Hence, as expected, oxygen was implicated in the deterioration of the chilling-sensitive tissue. The antioxidants had a more limited capacity in the cucumber to withstand the increased oxidative stress that accompanied the lower temperatures than did the chilling-tolerant pea leaf.

In a study of antioxidant enzymes, Jahnke et al.[48] reported a greater increase (60%) in monodehydroascorbate reductase activity in the chilling-susceptible *Zea mays* compared to the chilling-tolerant *Zea diploperennis* plants that were transferred to low temperatures. Of those investigated, only this enzyme might affect vitamin E metabolism directly (Figure 4A). More enzyme activity in the chilling-susceptible cultivar may accommodate an increased demand for metabolizing free radicals through vitamin E turnover. Although no data were provided, it is necessary to know if vitamin E content decreased to a greater extent in the chilling-susceptible variety in response to treatment at lower temperature.

Among the many data from the *Picea* species that were analyzed by Franzen et al.[18] are the interesting comparisons between current-year needle content of vitamin E and that of older needle tissue. Typically, the current-year tissue had α-tocopherol content less than 10 μg per gfw. The exact time of harvest after bud break was not indicated. However, following the first period of hardening, the concentration increased (80 to 100 μg per gfw), depending on the species, as noted for three examples listed in Table 1B. A significant increase in antioxidant concentration for vitamin C and glutathione, as well as for glutathione reductase, has also been correlated with the onset of hardening in Eastern white pine.[58] Hence, it may be that vitamin E synthesis is another metabolic event of the series that accompanies cold acclimation in conifers.

C. STORAGE

Antioxidant content of stored potato tubers was evaluated during a 40-week storage period. Spychalla and Desborough[16] anticipated an increase in oxidative stress under the storage conditions of 3 to 10°C. They observed a linear increase in vitamin E content up to fourfold by the fortieth week of storage. They also observed a parallel increase in catalase and an increase in superoxide dismutase (SOD) activity that was maximum by the fifth week in tubers stored at 3°C. Cultivars with the greater stability during storage, as monitored by quantifying ion leakage from tissue, had the higher levels of catalase and SOD and lower levels of vitamin E. The cultivar with the most damaged (leaky) membranes had the highest vitamin E content. As with the experiments of Wise and Naylor,[57] the increase in vitamin E may be responding only to increased lipid peroxidation rather than to some other regulatory signal.

These studies also revealed the importance of quantifying the cellular distribution of the antioxidants. It may be that the ion leakage assay preferentially monitors effects on the plasma membrane. Whether or not the plasma membrane contains significant quantities of vitamin E is not known, so that the relationship of vitamin E content to potato-tuber stability remains undefined. A quantitative profile of vitamin C content would be informative. This model may allow further quantitative examination of the relative contributions of antioxidants to overall tissue stability.

Vitamin E content has been quantified in another nonphotosynthetic tissue, barley seed, in order to define the influence of moisture content and atmospheric composition during storage. At 23% moisture, Hakkarainen et al. reported that vitamin E content decreased from 92 μg per gdw to 20 μg per gdw after 10 months of storage.[11] Greater loss of vitamin E occurred at 28% moisture. In barley, which contains significant quantities of the tocotrienols, there was actually a relative increase in the amount of vitamin E as a percentage of the total tocols. This change was greatest under those conditions that resulted in the greatest loss in total vitamin E. Furthermore, if the seed was stored with a source of external CO_2 supply, the decrease in vitamin E was less. These changes in the barley seed, after ripening processes have occurred, may be similar to the pregrowth changes that accompany seeds during germination. The extent to which any of these changes were affected by oxygen was not quantified, and no studies correlate the seed content of vitamin E with viability or germination frequency. The data, however, implicated the tocotrienols as potential sources of vitamin E and/or as having antioxidant function themselves.

D. NUTRIENT EFFECTS

In studies with suspension cell cultures of safflower, Furuya et al.[19] established the direct increase in vitamin E in response to providing phytol in the growth medium of the culture. The advantage of this response is the ability to increase production of the stereospecifically pure **d**-α-tocopherol. Ability to manipulate levels of an antioxidant like vitamin E provides opportunity to determine how vitamin E may affect the tolerance of cells to specific environmental stressors, e.g., colder temperatures and ozone exposure.

E. DROUGHT/ENVIRONMENT

Price and Hendry[59] quantified changes in malondialdehyde (MDA) and tocopherols in six plant species that were exposed to episodes of drought. They observed a parallel increase in MDA to that in vitamin E, and concluded that vitamin E biosynthetic capacity increases readily in response to the demands of oxidative stress associated with drought. Since no significant changes in SOD, catalase, and peroxidase were generally observed, it may be that vitamin E responded as a first line of defense to oxidative assault. Only in the native grass, *Deschampsia flexuosa* L., a slow growing, stress-tolerant species, did they observe stability of protein content and significant increases in SOD, catalase, and peroxidase activities, as well as the doubling of vitamin E content. The drought-sensitive grass, *Poa trivialis* L., responded with a threefold increase in MDA, but had only a low content of vitamin E. Further comparative studies of these grass species seem warranted in order to establish functional relationships between antioxidant metabolism and drought tolerance.

Kandil et al.[12] compared field plantings of sunflower hybrids in two locations, Germany and Giza, Egypt. Adequate moisture and fertilizer was provided at both sites, so that, generally, differences in growth temperature and ambient radiation were compared as they affected seed protein, oil quality, and tocopherol content. It was apparent that differences among cultivars was not significant; however, a reciprocal relationship existed between the partitioning of carbon between protein and oil content. The lowest protein and highest oil content occurred in the seed from plants grown at the German locations. The seed grown during the winter season in Egypt had similar oil quality (90% unsaturated fatty acids) to that grown in Germany, but the 950 ppm tocopherol in the Egypt-grown seed was significantly higher than that in the seed from Germany (860 ppm). No data were provided about leaf concentrations of vitamin E in this study. Relationships between carbon partitioning and sites of tocopherol synthesis need to be described in order to understand the physiological significance of responses to environmental factors. In experiments with barley seed, Hakkarainen et al.[11] reported that, in good growing seasons, vitamin E content varied between 80 to 100 μg per gdw. When the growing season was poor, e.g., in 1980, the vitamin E in barley was 55 to 65 μg per gdw. These results establish that quantitative differences in tocopherol content can be determined by environmental parameters.

F. HERBICIDES

Finckh and Kunert[60] used the diphenylether herbicide, oxyfluorfen [2-chloro-1-(3-ethoxy-4-nitrophenoxy)-4-(trifluoromethyl)benzene] to induce peroxidative damage in several plant species. Ethane was used to monitor cell damage. It was concluded that both the amount and the ratio of vitamin E and vitamin C contributed to the relative sensitivity of a seedling to the diphenylether. The greatest tolerance was reported for mustard (*Sinapis alba*) and sicklepod (*Cassia obtusifolia*) that had C to E ratios of 10 to 15 and 500 to 600 μg vitamin E per gdw. Jimson weed (*Datura stramonium*), on the other hand, was very susceptible to the herbicide, and had a vitamin C to E ratio of 1.2, but contained 830 μg vitamin E per gdw. No simple correlation existed between vitamin E content and sensitivity to the herbicide. There were no reported quantitative measurements of either vitamin after treatment with the herbicide, and the authors noted that other aspects of the antioxidant systems in plants may also play a role in protection against this herbicide. Studies with model plant systems and their response to herbicide treatments will be useful in establishing the role of vitamin E in the hierarchy of antioxidant protective mechanisms in plants.

VI. SUMMARY

Although our understanding of vitamin E function in plants is far from being complete, one may conclude the following: (1) particularly, in plants, vitamin E functions mainly as an antioxidant to quench organic peroxyl radicals in the lipid phase of membranes; (2) the phytyl chain of vitamin E contributes to membrane stability independent of the antioxidant properties of the chromanol ring; (3) increase in vitamin E content in tissues responds to increased lipid peroxidation; (4) in concert with vitamin C and possibly with GSH, there is capacity for reduction of the tocopheroxyl radical to the tocopherol, and this cycling accounts for the synergistic protection afforded by these lipophilic and hydrophilic compounds; and (5) no specific regulatory role or function has been established for vitamin E.

Many questions remain unanswered. No experiments identify a specific regulatory role for vitamin E. The absence of definitive quantitative data on the distribution of the tocopherols in plant membranes allows only ambiguous interpretations of responses in vitamin E content to stress. At this time, it is possible only to speculate how tocopherol synthesis is regulated and what signals direct this regulation. The specificity for high concentrations of the tocotrienols in certain plant tissues is unexplained, but they may reflect the greater availability of geranylgeranol derivatives compared to phytol. The potential uses of mutations and cell/tissue cultures to explore the function as well as the regulation of vitamin E synthesis seem promising. Model plant systems that allow careful control of environmental parameters are becoming better described, and their use should provide a clearer understanding of vitamin E function in plant metabolism.

REFERENCES

1. **Pennock, J. F., Hemming, F. W., and Kerr, J. D.,** A reassessment of tocopherol chemistry, *Biochem. Biophys. Res. Commun.*, 17, 542, 1964.
2. **Janiszowska, W. and Pennock, J. F.,** The biochemistry of vitamin E in plants, in *Vitamins and Hormones: Advances in Research and Applications,* Munson, P. L., Glover, J., Diczfauly, E., and Olson, R. E., Eds., vol. 34, Academic Press, New York, 1976, 77.
3. **Diplock, A. T., Machlin, L. J., Packer, L., and Pryor, W. A.,** Eds., Vitamin E: biochemistry and health implications, *Ann. N.Y. Acad. Sci.*, 570, 555p, 1989.
4. **Diplock, A. T.,** Chairperson, Biology of vitamin E, *Ciba Foundation Symposium 101,* Pitman Books, London, 1983, 260.
5. **Pryor, W. A.,** Can vitamin E protect humans against the pathological effects of ozone in smog?, *Am. J. Clin. Nutr.*, 53, 702, 1991.
6. **Battle, R. W., Gaunt, J. K., and Laidman, D. L.,** The effect of photoperiod on endogenous γ-tocopherol and plastochromanol in leaves of *Xanthium strumarisum* L. (Cocklebur), *Biochem. Soc. Trans.*, 4, 484, 1976.
7. **Burton, G. W. and Ingold, K. U.,** Vitamin E as an *in vitro* and *in vivo* antioxidant, in *Vitamin E: Biochemistry and Health Implications,* Diplock, A. T., Machlin, L. J., Packer, L., and Pryor, W. A., Eds., *Ann. N.Y. Acad. Sci.*, 570, 7, 1989.
8. **Gomez-Fernandes, J. C., Villalain, J., Aranda, F. J., Oritz, A., Micol, V., Coutinho, A., Berberan-Santos, M. N., and Prieto, M. J. E.,** Localization of α-tocopherol in membranes, in *Vitamin E: Biochemistry and Health Implications,* Diplock, A. T., Machlin, L. J., Packer, L., and Pryor, W. A., Eds., *Ann. N.Y. Acad. Sci.*, 570, 109, 1989.
9. **Larson, R. A.,** The antioxidants of higher plants, *Phytochemistry,* 27, 969, 1988.
10. **Green, J.,** The distribution of tocopherols during the life-cycle of some plants, *J. Sci. Food Agric.*, 9, 801, 1958.
11. **Hakkarainen, R. V. J., Tyopponen, J. T., and Bengtsson, S. G.,** Relative and quantitative changes in total vitamin E and isomer content of barley during conventional and airtight storage with special reference to annual variations, *Acta Ag. Scand.*, 33, 395, 1983.

12. **Kandil, A., Ibrahim, A. F., Marquard, R., and Taha, R. S.,** Response of some quality traits of sunflower seeds and oil to different environments, *J. Agro. Crop Sci.*, 164, 224, 1990.
13. **Schultze, G.,** Biosynthesis of α-tocopherol in chloroplasts of higher plants, *Fat Sci. Technol.*, 92, 86, 1990.
14. **Schultze, G., Heintze, A., Hoppe, P., Hagelstein, P., Gorlach, J., Meereis, K., Schwanke, U., and Preiss, T.,** Tocopherol and carotenoid synthesis in chloroplasts. Its tight linkage to plastidic carbon metabolism in developing chloroplasts, in *Active Oxygen/Oxidative Stress and Plant Metabolism*, Pell, E. and Steffan, K. L., Eds., *Curr. Topics in Plant Physiol.*, 6, 156, 1991.
15. **Yerin, A. N., Kormanovskii, A. Ya., and Ivanov, I. I.,** Localization of α-tocopherol in chloroplasts, *Biofizika*, 29, 363, 1984.
16. **Spychalla, J. P. and Desborough, S. L.,** Superoxide dismutase, catalase, and α-tocopherol content of stored potato tubers, *Plant Physiol.*, 94, 1214, 1990.
17. **Gapor, Ab., Kato, A., and Ong, A. S. H.,** α-tocopherol content in oil palm leaflet, *J. Am. Oil Chem. Soc.*, 63, 330, 1986.
18. **Franzen, J., Bausch, J., Glatzle, D., and Wagner, E.,** Distribution of vitamin E in spruce seedling and mature tree organs, and within the genus, *Phytochemistry*, 30(1), 147, 1991.
19. **Furuya, T., Yoshikawa, T., Kimura, T., and Kaneko, H.,** Production of tocopherols by cell culture of safflower, *Phytochemistry*, 26, 2741, 1987.
20. **Wellburn, A. R.,** Studies on the biosynthesis of the tocopherols in higher plants, *Phytochemistry*, 9, 743, 1970.
21. **Marshall, P. S., Morris, S. R., and Threlfall, D. R.,** Biosynthesis of tocopherols: a reexamination of the synthesis and metabolism of 2-methyl-6-phytyl-1,4-benzoquinol, *Phytochemistry*, 24, 1705, 1985.
22. **Pennock, J. F.,** The biosynthesis of chloroplastidic terpenoid quinones and chromanols, *Biochem. Soc. Trans.*, 11, 504, 1983.
23. **Soll, J., Schultz, G., Rudiger, W., and Benz, J.,** Hydrogenation of geranylgeraniol: two pathways exist in spinach chloroplasts, *Plant Physiol.*, 71, 849, 1983.
24. **Schmidt, D. L., Grundemann, D., Groth, G., Muller, B., Hennig, H., and Schultz, G.,** Shikimate pathway in non-photosynthetic tissues. Identification of common enzymes and partial purification of dehydroquinate hydrolase-shikimate oxidoreductase and chorismate mutase from roots, *J. Plant Physiol.*, 138, 51, 1991.
25. **Rise, M., Cojocaru, M., Gottlieb, H. E., and Goldschmidt, E. E.,** Accumulation of α-tocopherol in senescing organs as related to chlorophyll degradation, *Plant Physiol.*, 89, 1028, 1989.
26. **Henry, A., Powls, R., and Pennock, J.,** *Scenedesmus obliquus* PS28: a tocopherol-free mutant which cannot form phytol, *Biochem. Soc. Trans.*, 14, 958, 1986.
27. **Dunphy, P. J., Whittle, K. J., Pennock, J. F., and Morton, R. A.,** Identification and estimation of tocotrienols in *Hevea* latex, *Nature*, 207, 521, 1965.
28. **d'Harlingue, A. and Camara, B.,** Plastid enzymes of terpenoid biosynthesis: purification and characterization of γ-tocopherol methyltransferase from *Capsicum* chromoplasts, *J. Biol. Chem.*, 260, 15200, 1985.
29. **Janiszowska, W.,** Intracellular localization of tocopherol biosynthesis in *Calendula officinalis*, *Phytochemistry*, 26, 1403, 1987.
30. **Thomas, D. R. and Stobart, A. V.,** Quinones and α-tocopherol in greening callus cultures of *Kalanchoe crenata*, *New Phytol.*, 70, 163, 1971.
31. **Heintze, A., Horlach, J., Liuschner, C., Hoppe, P., Hagelstein, P., Schulze-Siebert, D., and Schultz, G.,** Plastidic isoprenoid synthesis during chloroplast development, *Plant Physiol.*, 93, 1121, 1990.
32. **Threlfall, D. R.,** The biosynthesis of vitamins E and K and related compounds, *Vitam. Horm.*, 29, 153, 1971.
33. **Kagan, V. E.,** Tocopherol stabilizes membrane against phospholipase A, free fatty acids, and lysophospholipids, in *Vitamin E: Biochemistry and Health Implications*, Diplock, A. T., Machlin, L. J., Packer, L., and Pryor, W. A., Eds., *Ann. N.Y. Acad. Sci.*, 570, 121, 1989.

34. **Erin, A. N., Skrypin, V. V., and Kagan, V. E.,** Formation of α-tocopherol complexes with fatty acids. Nature of complexes, *Biochim. Biophys. Acta,* 815, 209, 1985.
35. **Gorbunov, N. V., Kagan, V. E., Aiekseev, S. M., and Erin, A. N.,** Role of the isoprenoid chain of lateral mobility of α-tocopherol in the lipid bilayer, *Bull. Exp. Biol. Med.,* 112, 946, 1988.
36. **Baszynski, T.,** Effect of α-tocopherol on reconstitution of PSI in heptane-extracted spinach chloroplasts, *Biochim. Biophys. Acta,* 347, 31, 1974.
37. **Mascio, P., Devasagayam, T. P. A., Kaiser, S., and Sies, H.,** Carotenoids, tocopherols, and thiols as biological singlet molecular oxygen quenchers, *Biochem. Soc. Trans.,* 18, 1054, 1990.
38. **Van Hasselt, P. R., De Kok, L. J., and Kuiper, P. J.,** Effect of α-tocopherol, β carotene, monogalactosyldiglyceride and phosphatidylcholine on light-induced degradation of chlorophyll in acetone, *Physiol. Plant,* 45, 475, 1979.
39. **Skinner, W. A. and Parkhurst, R. M.,** Antioxidant properties of α-tocopherol derivatives and relationship of antioxidant activity to biological activity, *Lipids,* 5, 184, 1970.
40. **Burton, G. W., Cheeseman, K. H., Doba, T., Ingold, K. R., and Slater, T. F.,** Vitamin E as an antioxidant *in vitro* and *in vivo,* in Diplock, A. T., Chairperson, Biology of vitamin E, *Ciba Foundation Symposium 101,* Pitman Books, London, 1983, 4.
41. **Yagi, K., Yamada, H., and Nishikimi, M.,** Oxidation of α-tocopherol with O_2^-, in Tocopherol, Oxygen and Biomembranes, deDuve, S. and Hayaishi, O., Eds., Elsevier, Amsterdam, 1978, 1.
42. **Doba, T., Burton, G. W., and Ingold, K. U.,** Antioxidant and co-antioxidant effect of vitamin C. The effect of vitamin C, either alone or in the presence of vitamin E or a water-soluble vitamin E analog, upon the peroxidation of aqueous multilamellar phospholipid liposomes, *Biochim. Biophys. Acta,* 835, 298, 1985.
43. **Niki, E., Kawakami, A., Yamamoto, Y., and Kamiya, Y.,** Oxidation of lipids. VIII. Synergistic inhibition of oxidation of phosphatidylcholine liposome in aqueous dispersion by vitamin E and vitamin C, *Bull. Chem. Soc. Jpn.,* 58, 1971, 1985.
44. **Packer, J. E., Slater, T. F., and Willson, R. L.,** Direct observation of a free radical interaction between vitamin E and vitamin C, *Nature,* 278, 737, 1979.
45. **Niki, E., Saito, T., Kawakami, A., and Kamiya, Y.,** Inhibition of oxidation of methyl linoleate in solution between vitamin E and vitamin C, *J. Biol. Chem.,* 259, 4177, 1984.
46. **Hossain, M. A., Nakano, Y., and Asada, K.,** Monodehydroascorbate reductase in spinach chloroplasts and its participation in regeneration of ascorbate for scavenging hydrogen peroxide, *Plant and Cell Physiol.,* 25, 385, 1984.
47. **Polle, A., Chakrabarti, Kl, Schurmann, W., and Rennenberg, H.,** Composition and properties of hydrogen peroxide decomposing systems in extracellular and total extracts from needles of Norway spruce (*Picea abies* L., karst), *Plant Physiol.,* 94, 312, 1990.
48. **Jahnke, L. S., Hull, M. R., and Long, S. P.,** Chilling stress and oxygen metabolizing enzymes in *Zea mays* and *Zea diploperennis, Plant Cell and Environ.,* 14, 97, 1991.
49. **Hossain, M. A. and Asada, K.,** Monodehydroascorbate reductase from cucumber is a flavin adenine dinucleotide enzyme, *J. Biol. Chem.,* 260, 12920, 1985.
50. **McCay, P. B., Brueggemann, G., Lai, E. K., and Powell, S. R.,** Evidence that α-tocopherol functions cyclically to quench free radicals in hepatic microsomes: requirement for glutathione and a heat-labile factor, in *Vitamin E: Biochemistry and Health Implications,* Diplock, A. T., Machlin, L. J., Packer, L., and Pryor, W. A., Eds., *Ann. N.Y. Acad. Sci.,* 570, 32, 1989.
51. **Niki, E., Tsuchiya, J., Tanimura, R., and Kamiya, Y.,** Regeneration of vitamin E from α-chromanoxyl radical by glutathione and vitamin C, *Chem. Lett.,* 1982, 789, 1982.
52. **Goin, J., Gibson, D. D., McCay, P. B., and Cadenas, E.,** Glutathionyl- and hydroxyl radical formation coupled to the redox transitions of 1,4-naphthoquinone bioreductive alkylating agents during two-electron reductive addition, *Arch. Biochem. Biochem.,* 288, 386, 1991.
53. **Booth, V. H.,** The rise in tocopherol content in wilting and in non-illuminated leaves, *Phytochemistry,* 3, 273, 1964.
54. **Hess, J. L., Pallansch, M. A., Harich, K., and Bunce, G. E.,** Quantitative determination of α-tocopherol on thin layers of silica gel, *Anal. Biochem.,* 83, 401, 1977.

55. **Kunert, K. J. and Ederer, M.,** Leaf aging and lipid peroxidation: the role of antioxidants vitamin C and E, *Physiol. Plant,* 65, 85, 1985.
56. **Alberdi, M. and Corcuera, L. J.,** Cold acclimation in plants, *Phytochemistry,* 30, 3177, 1991.
57. **Wise, R. R. and Naylor, A. W.,** Chilling-enhanced photoxidation: evidence for the role of singlet oxygen and superoxide in the breakdown of pigments and endogenous antioxidants, *Plant Physiol.,* 83, 278, 1987.
58. **Anderson, J. A., Chevone, B. I., and Hess, J. L.,** Seasonal variation in the antioxidant system of Eastern white pine needles: evidence for thermal dependence, *Plant Physiol.,* 98, 501, 1992.
59. **Price, A. and Hendry, G.,** The significance of the tocopherols in stress survival in plants, in *Free Radicals, Oxidant Stress and Drug Action,* Rice-Evans, C., Ed., Richelieu Press, London, 1987, 443.
60. **Finckh, B. F. and Kunert, K. J.,** Vitamins C and E: an antioxidative system against herbicide-induced lipid peroxidation in higher plants, *J. Agric. Food Chem.,* 33, 574, 1985.
61. **Bochkaryou, N. I.,** All Union Research Institute of Oil Crops, Krasnodar, Russia, personal communication.

Chapter 6

PLANT PHENOLICS

Norman G. Lewis

TABLE OF CONTENTS

0-8493-6328-4/93/$0.00 + $.50

I. OVERVIEW

Terrestrial vascular plants synthesize a structurally, biogenetically diverse array of phenolic products that are compartmentalized or accumulate in specific tissues or organs. Most phenols readily undergo oxidation to form colored, quinone-containing products when exposed to air; this response is frequently observed as "browning" reactions of plant tissue as part of a healing response to wounding or invasion by opportunistic pathogens. A familiar example of this is the rapid discoloration of wounded potato tubers. Consequently, many of these metabolites or their derivatives find industrial application as antioxidants, e.g., in food preparations, by minimizing the effects of oxidative reactions of other components leading to discoloration and spoil. Thus, plant phenolics have the potential to function as antioxidants by trapping free radicals generated in oxidative chemistry which then normally undergo coupling reactions leading eventually to (colored) polymeric or oligomeric products.

The most common classes of plant phenolics having antioxidant properties include those metabolites derived from the so-called hydrolyzable gallo- and ellagitannins; phenylpropanoids — including, among others, lignins, hydroxycinnamic acids and their derivatives, coumarins, monolignols, lignans and neolignans, and products of the phenylpropanoid-acetate pathway — alkyl ferulates; suberins; flavonoids; and other mixed pathway metabolites, e.g., furanocoumarins. Many of these antioxidants are stored in "sequestered" form until needed (e.g., some metabolites are known to be stored in the vacuole, or are present in the cell wall), whereas others are synthesized *de novo* in response to pathogen invasion at or near the point of attack. The polymers formed from plant phenolics in the cell wall provide structural support and form barriers to prevent moisture loss/diffusion and pathogen encroachment. The phenols also function in defense mechanisms that depend on UV protectant, antifungal, antiviral, antibacterial, antifeedant, and antimitotic properties, and also function in signal induction and transduction and in morphogenesis.

This chapter is restricted to phenolic antioxidants present in vascular plants, and not those in aquatic plants that appear to be mainly acetate-derived. The analysis is by no means fully comprehensive, but instead, focuses on some of the major classes of phenolics found in vascular plants, with exception of the flavonoids which have been described comprehensively elsewhere. The first four sections of the chapter describe what is currently understood about the biosynthetic pathways of phenolic antioxidants, in terms of enzymes, intermediates, and their compartmentation. The last section summarizes our understanding regarding known or possible functions and their contemplated roles as antioxidants in higher plants.

II. PHENOLS

A. GALLIC AND PROTOCATECHUIC ACIDS

Gallic acid **1** is most typically found as a constituent of polymeric gallo- and ellagitannins; however, it can also occur conjugated to miscellaneous sugars, glycosides, polyols, and polyphenols,[1] as illustrated by the galloyl esters **2–4** that are linked to shikimic acid,[2] glycerol,[2] and (−)-epigallocatechin,[3] respectively.

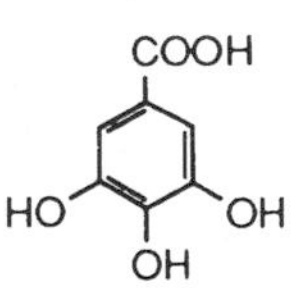

Gallic acid **1**

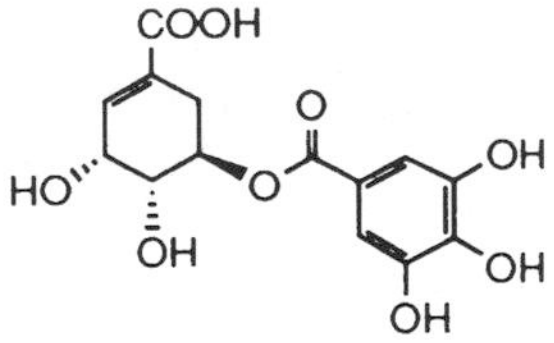

3-*O*-galloyl shikimic acid **2**

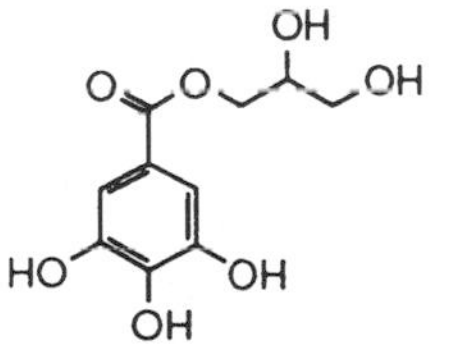

Galloyl glycerol **3**

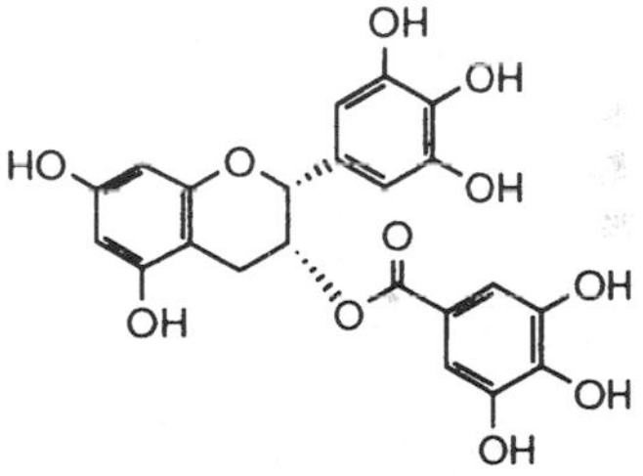

Epigallocatechin gallate **4**

For such an apparently simple phenolic compound, it is quite surprising that neither the enzymology of gallic acid **1** formation nor the subcellular location of its synthesis and storage are defined. Three biosynthetic routes to gallic acid **1** have been proposed based upon radiolabeling and enzyme-inhibition studies.[4] These include: direct formation via dehydrogenation of dehydroshikimic acid **5** (from the shikimic acid **6** pathway),[4] or indirectly, via the so-called β-oxidation route from caffeic acid **7** (a phenylpropanoid pathway metabolite) via either protocatechuic **8** or 3,4,5-trihydroxycinnamic **9** acids, respectively (Figure 1).

Most evidence favors the direct route, based upon the following lines of evidence: (1) gallic acid **1** accumulation increased fivefold when *Quercus robur* suspension cultures were treated with glyphosate (*N*-[phosphonomethyl] glycine), an inhibitor of 5-enolpyruvyl shikimate-3-phosphate synthetase, and thus also of phenylalanine and caffeic acid **7** synthesis.[5] It is probable that gallic acid **1** accumulates because this enzymatic step is inhibited and results in reduced carbon flow into the pathways leading to the aromatic amino acids (phenylalanine, Phe, and tyrosine, Tyr), their metabolites, and, hence, the phenylpropanoids. A redistribution of carbon flow is then envisaged to occur resulting in increased gallic acid **1** synthesis; (2) when velvetleaf (*Abutilon theophrasti*) was treated with 5 m*M* glyphosate for 6 days, specific phenolics accumulated — gallic **1** (*ca.* twofold) and protocatechuic **8** (*ca.* tenfold) acids, respectively.[6] By contrast, the phenylpropanoid metabolites,

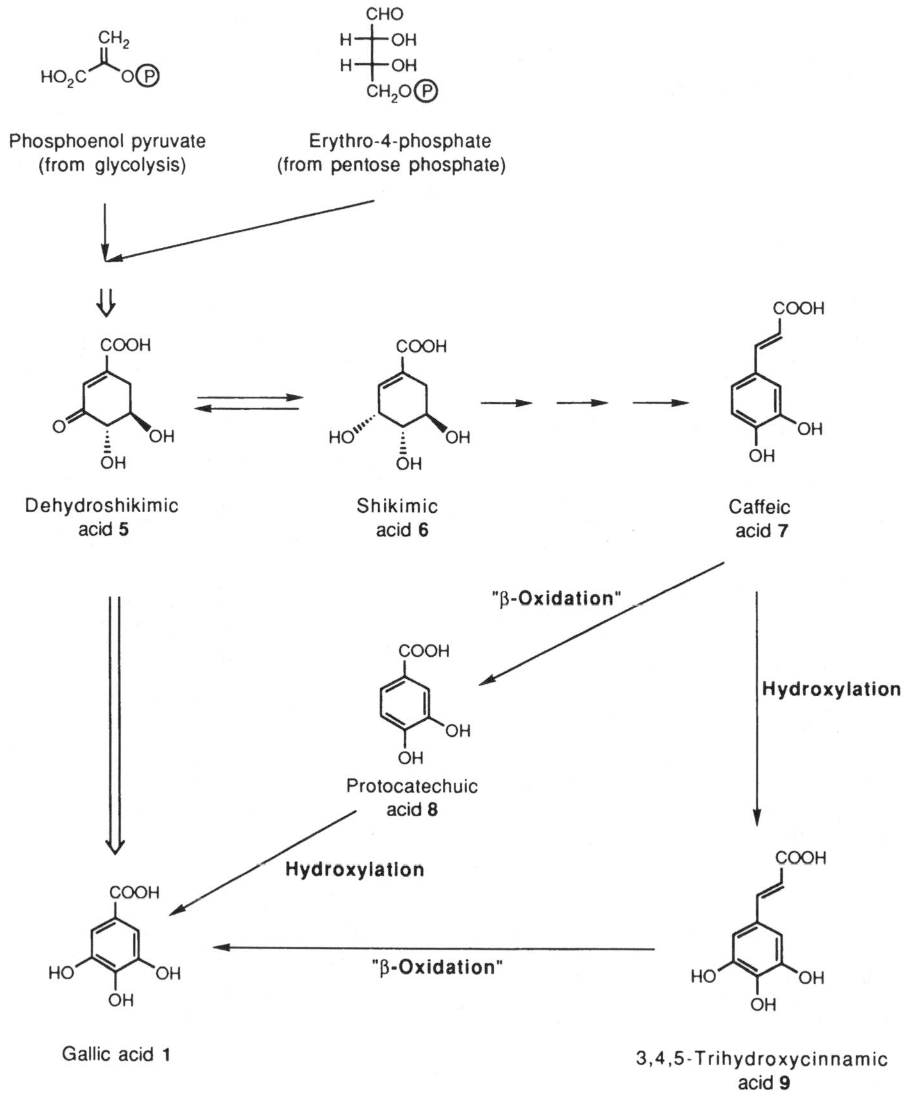

FIGURE 1. Proposed biosynthetic pathways to gallic acid **1**.

p-coumaric **10**, caffeic **7**, and ferulic **11** acids decreased in amounts by approximately 75, 56, and 39%; and (3) in the tea plant, (*Camellia sinensis*), [$^{14}CO_2H$] shikimic acid **6** was apparently incorporated into gallic acid **1** (1.3% incorporation).[7] Obviously, if either of the two indirect "β-oxidation" routes from caffeic acid **7** to gallic acid **1** were operative, the radiolabeled carboxyl group would have been lost during phenylalanine (or tyrosine) biosynthesis.

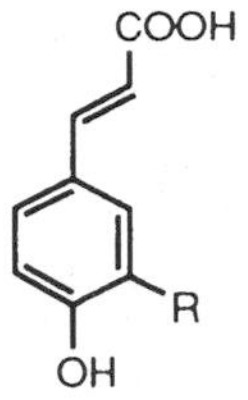

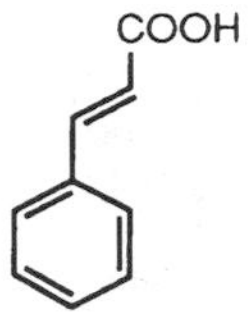

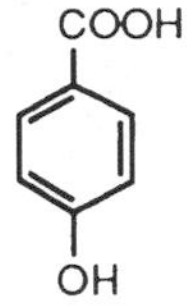

R=H : *p*-Coumaric acid **10**
R=OCH_3 : Ferulic acid **11**

Cinnamic acid **12**

p-Hydroxybenzoic acid **13**

Taken together, these results strongly suggest that, at least in the species examined, the indirect pathways utilizing either caffeic **7** or 3,4,5-trihydroxycinnamic **9** acids are not required for gallic **1** and protocatechuic **8** acid formation. However, it must be cautioned that these investigations do not unequivocally rule out "β-oxidation" as a minor pathway, or as an alternate route, in other plants. Evidence for the indirect pathway is as follows: (1) [3 – ^{14}C] D,L-phenylalanine was metabolized into [^{14}C] gallic acid **1** in *Rhus typhina*,[8] where it was proposed that 3,4,5 trihydroxycinnamic acid **9** served as a putative intermediate. Note, though, that this acid has never been demonstrated to occur as a natural product; and (2) [3- – ^{14}C] cinnamic acid **12** was converted into protocatechuic acid **8** in *Hydrangea macrophylla* and *Gaultheria procumbens,* as was [3 – ^{14}C] caffeic acid **7** in *Orzyva sativa*. Additionally, [$^{14}CO_2H$] protocatechuic acid **8** was apparently metabolized into gallic acid **1** in *Pelargonium hortorum*.[9]

Thus, a body of evidence in other plant species exists in support of conversion of phenylpropanoid metabolites, such as caffeic acid **7** into protocatechuic **8** and gallic **1** acids. Additional support for an indirect pathway comes from studies using *Lithospermum erythrorhizon* cell cultures, where it was recently established that *p*-hydroxybenzoic acid **13** was derived from *p*-coumaric acid **10**.[10] This transformation was proposed to proceed via a "nonoxidative" pathway rather than by "β-oxidation", although the enzymes were not identified. A similar situation might be envisaged for gallic **1** and protocatechuic **8** acids. Therefore, in light of such findings, there is a need to clarify the relative importance of both direct and indirect, i.e., "β-oxidation" or "nonoxidative", routes in protocatechuic **8** and gallic **1** acid biogenesis. This understanding will require precise definition of enzymology, and the subcellular location of enzymes and metabolites.

B. GALLO- AND ELLAGITANNINS

"Hydrolysable tannins" are of limited distribution in nature, being found in leaves, fruits, pods, and galls (and in some cases, in wood and bark) of dicots, such as oak, chestnut, sumac, and others;[1,4] they have not been found in monocots.[11] The general term "hydrolysable" simply reflects the sensitivity of such substances to hydrolytic agents and tannases (esterases), whereas "tannins" derive their name from previous extensive usage as leather-tanning agents. "Hydrolysable tannins" range in complexity from simple esters, such as β-glucogallin **14**, to polyesters

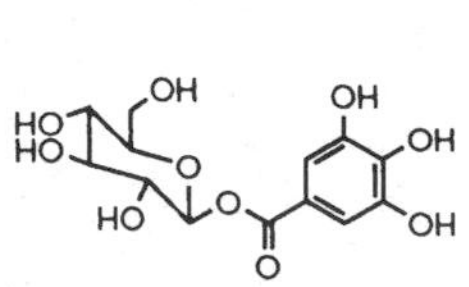

β-Glucogallin **14**

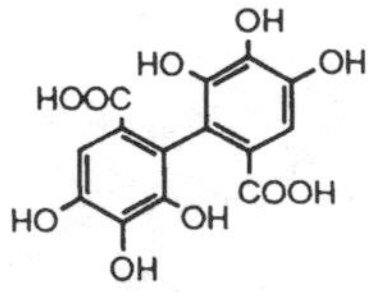

Hexahydroxydiphenic acid **15** (= G-G)

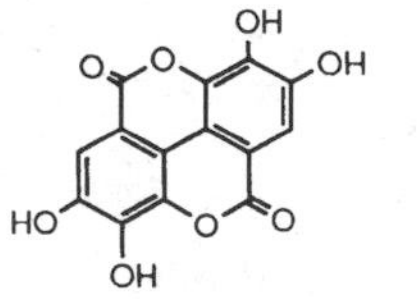

Ellagic acid **16**

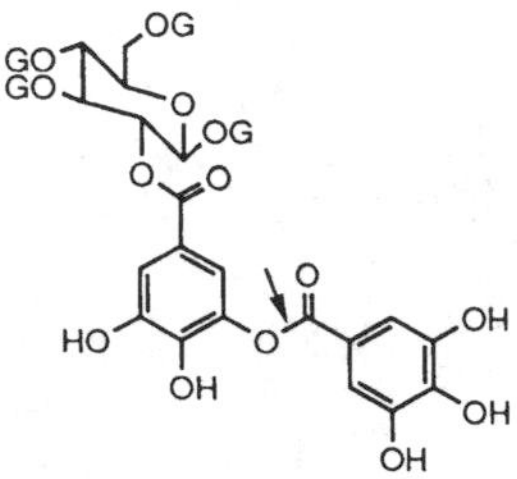

→ = depside linkage (Type A)

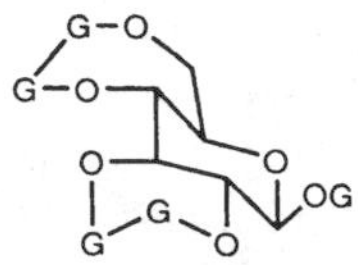

Hexahydroxydiphenoyl (Type B)

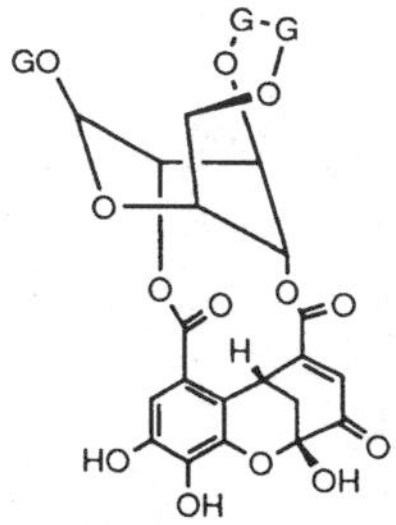

Hexahydroxydiphenoyl and dehydrohexahydroxydiphenoyl (Type C)

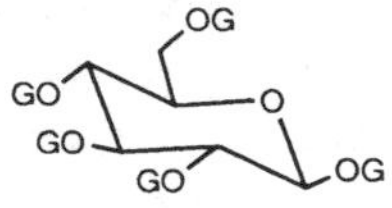

β-Glucopentagallin **17**

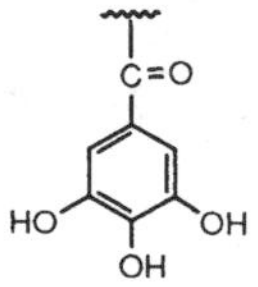

G = Galloyl residue

FIGURE 2. Typical bonding environments of gallic acid (derivatives) in higher plants.

having a polyol core (normally glucose) with esterified galloyl **1** or hexahydroxydiphenoyl **15** residues, or derivatives thereof. A recent survey[1] of "hydrolysable tannins" in leaf tissue of various dicots indicated that they can be classified into three broad types, i.e., those bonded via depside (Type A), hexahydroxydiphenoyl (Type B), or hexahydroxydiphenoyl/dehydrohexahydroxydiphenoyl (Type C) linkages (Figure 2); significantly, only one bonding type is normally present in a particular species. Molecular weight ranges reported for various soluble hydrolysable tannin preparations may not be particularly instructive, since, on aging in woody plant tissue (e.g., as for the ellagitannins of *Castanea sativa* and *Quercus petraea*), the tannin fraction becomes progressively "insoluble", due to presumed nonenzymatically driven reactions that result in their attachment to cell-wall components.[12]

Investigations of the biogenesis of gallo- and ellagitannins have largely focused upon the enzymology accompanying β-glucopentagallin **17** formation, using cell-free preparations from oak (*Quercus robur*) leaves,[13] which also synthesize Type

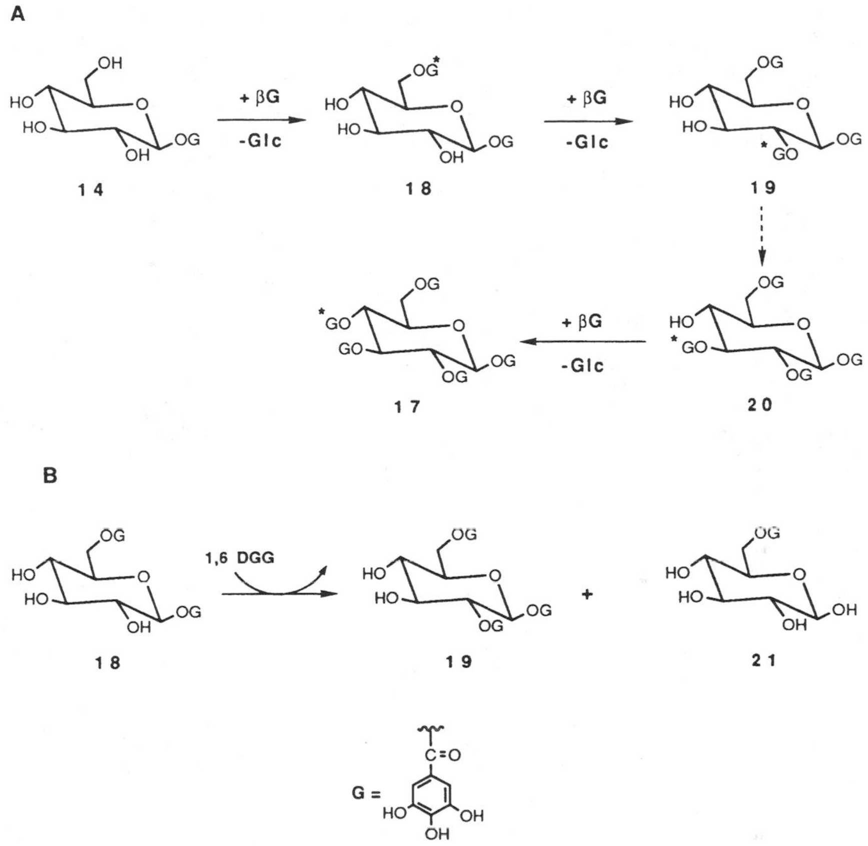

FIGURE 3. Proposed biosynthetic routes to (a) β-glucopentagallin **17** in *Quercus robur* and (b) 1,2,6-trigalloyl glucose **19** in *Rhus typhina*; +βG designates addition of gallic acid unit at position marked with *; − Glc refers to loss of glucose unit with each reaction.

B ellagitannins. With this biological system, the first enzymatic step in gallotannin biogenesis involves formation of β-glucogallin **14**, using both gallic acid **1** and UDP-D glucose as substrates. The β-glucogallin **14** then undergoes an enzymatically catalyzed "disproportionation" to form 1,6-digalloylglucose **18** (Figure 3A). Additional galloyl moieties are added at C-2, C-3, and C-4, again using β-glucogallin **14** as a co-substrate and galloyl donor. Although not yet unambiguously established, it is expected that specific enzymes are required for each galloylation step.

But this apparently logical sequence of events became complicated recently with results from investigations using sumac (*Rhus typhina*), which produces both β-glucopentagallin **17** and Type A gallotannins.[14] In this species, in addition to the route described above, 1,2,6-trigalloyl glucose **19** formation can also be catalyzed by another enzyme which "disproportionates" 1,6-digalloyl glucose **18** into 1,2,6-trigalloyl glucose **19** and 6-galloyl glucose **21**, respectively (Figure 3B); this particular enzyme was purified 1700-fold, had a molecular weight of 56 kD, and a Km = 11.5 m*M*. While the significance of such competing enzymatic transformations awaits an explanation,[14] this type of observation may provide a clue to explaining the unusual galloylation sequences in other plants, such as observed for the galloyl esters of bergenin **22–24** and norbergenin **25** in *Mallotus japonicus*.[2]

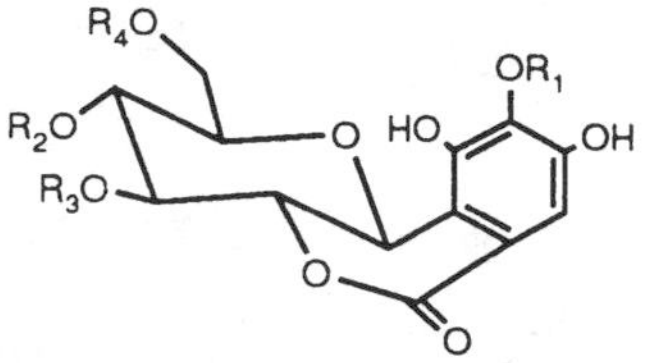

	R_1	R_2	R_3	R_4
22	Me	G	G	H
23	Me	H	G	G
24	Me	G	G	G
25	H	H	G	H

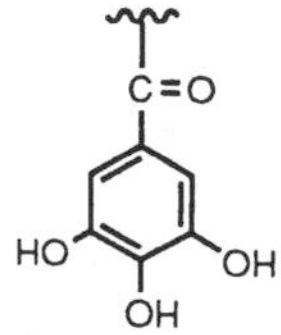

Gallotannin formation proper involves the addition of galloyl residues via depside linkages (Type A, Figure 2). Recent studies using *R. typhinus* cell-free extracts established that β-glucogallin **14** (and not galloyl-CoA) serves as galloyl donor during depside linkage formation, as shown by the sequential galloylation of β-glucopentagallin **17**.[15] Preliminary nuclear magnetic resonance (NMR) spectroscopic analyses of these enzymatic products suggested that new galloyl moieties were attached to galloyl residues linked at C_2, C_3, and C_4 of β-glucopentagallin **17**, a substitution pattern consistent with that of the polygalloylglucoses in *Rhus semialata*. It now needs to be determined whether one or more than one galloyl transferase is involved, and the order of galloylation needs to be established.

The next level of structural complexity in the "hydrolysable tannins" resides with the ellagitannins, e.g., hexahydroxydiphenic acid **15** (Type B, Figure 2), that are formed via 2,2′-coupling of galloyl **1** residues; this coupling is considered to occur only after attachment of the galloyl moieties to the glucose core. During isolation, hexahydroxydiphenic acid **15** spontaneously cyclizes to give the corresponding dilactone, ellagic acid **16**, and hence, these polyphenols are normally referred to as ellagitannins. Surprisingly, the enzyme catalyzing the (presumed) specific coupling of galloyl residues to give the ellagitannins has not been reported. Lastly, depending upon the species, the hexahydroxydiphenic components of the ellagitannins can undergo subsequent transformations involving aromatic ring fission following oxidative or reductive processes;[16] however, the biochemistry of such processes has not been investigated to date.

Given these data, it is apparent that, while exciting progress is being made, our understanding of gallo- and ellagitannin formation and regulation is only in its infancy. More findings in this area are eagerly awaited; the enzymology of β-glucopentagallin formation must be precisely established, as well as the mechanism of assembly of the complex (Type A–C) polyphenols. This paucity of knowledge severely limits the application of molecular biological techniques to the study of this branch of plant phenolic metabolism. None of the pathway-specific enzymes have been cloned, nor have any antibodies been raised against them to determine the subcellular location of enzymes by immunocytochemical means. Indeed, our detailed knowledge is so sparse that not even the subcellular compartmentalization

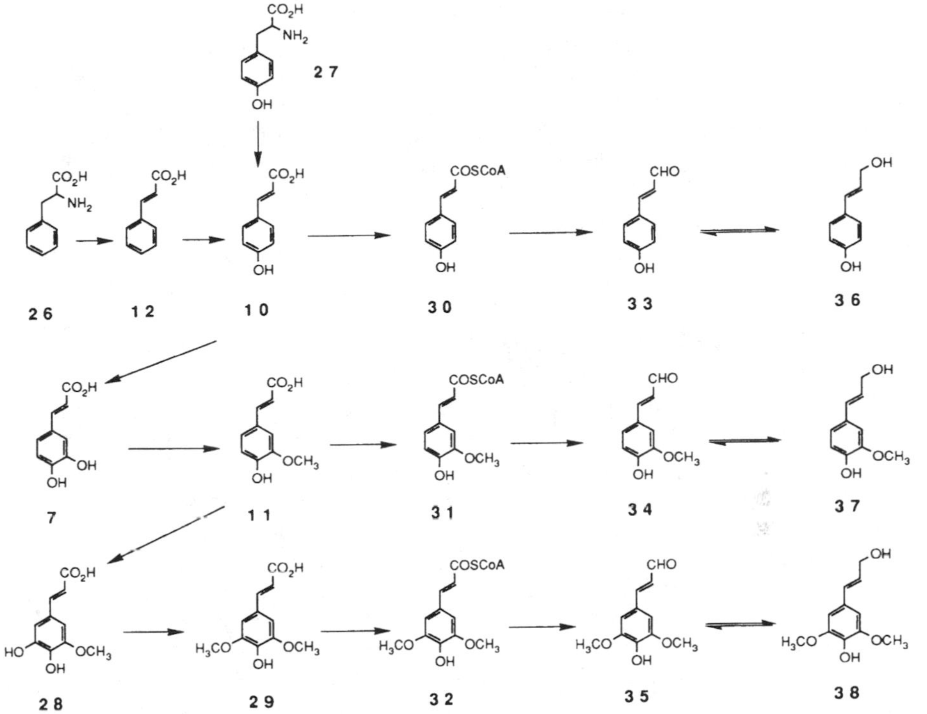

FIGURE 4. Elaboration of various phenylpropanoid metabolites from phenylalanine **26**/tyrosine **27**.

of the gallo- and ellagitannins has been determined. The evidence for vacuole accumulation is indirect; it is based upon our knowledge that proanthocyanidins (a flavonoid class of compounds) are present in the vacuole.

III. PHENYLPROPANOIDS

Phenylpropanoids **proper** are a diverse class of natural products derived exclusively from cinnamic **12** and *p*-coumaric **10** acids, or other metabolites originating **solely** from either compound. Entry into the phenylpropanoid pathway involves deamination of either phenylalanine, Phe **26** or, to a lesser extent in grasses and cereals, of tyrosine, Tyr **27**, to form *E*-cinnamic **12** and *p*-coumaric **10** acids, respectively (Figure 4). Next to the carbohydrates, they represent the second largest "sink" of carbon in vascular plants.[17] Occasionally, the term "general phenylpropanoid metabolism" is erroneously used to restrict the pathway to only those metabolic steps involving the deamination of Phe **26**/Tyr**27** and concluding with cinnamoyl CoA ester formation (for example, Reference 18). However, this "definition" is too restrictive, and a more accurate definition includes all metabolites *solely* derived from phenylpropanoid precursors/metabolites. Subclasses of metabolites involving only the phenylpropanoid pathway include the substituted hydroxycinnamic acids, monolignols, neolignans, lignans, and lignins, and simple coumarins, as well as *p*-hydroxybenzoic acids/aldehydes (formed by cleavage of phenylpropanoid precursors). (As discussed in the previous section, there is some evidence that gallic **1** and protocatechuic **8** acid-derived compounds may also partly be of phenylpropanoid origin.) Our current knowledge of the biochemistry/molecular biology is described below for each of the major enzymatic steps in phenylpropanoid metabolism. However, the pathways shown in Figure 4 represent a composite picture of studies carried out with many different plant types; no single plant has been thoroughly examined in terms of the enzymology and integrity of the enzyme-catalyzed steps in the pathway.

A. PRECURSORS AND AMMONIA LYASE

1. Phenylalanine (and Tyrosine) Ammonia Lyase

In dicots, phenylalanine ammonia lyase, (PAL) (E.C. 4.3.1.5) catalyzes the **anti**-elimination of ammonia from L-Phe **26** to give *E*-(*trans*)-cinnamic acid **12** (Figure 4).[19,20] The enzyme requires no co-factor for activity[21] and removes the pro-3S hydrogen from Phe **26**;[22,23] it does not use Tyr **27** as a substrate. By contrast, in grasses, in addition to Phe **26**, Tyr **27** can also undergo an enzymatically catalyzed deamination to yield *p*-coumaric acid **10**; this transformation is putatively catalyzed by a distinct tyrosine ammonia lyase (TAL).[24,25] It should be noted, however, that while in grasses both TAL and PAL activities have been observed, it has not been possible to obtain distinct enzymes capable of separately catalyzing *p*-coumaric **10** and cinnamic **12** acid formation, respectively.[21] One interpretation of such findings is that in grasses, PAL and TAL are one and the same, i.e., the enzyme is simply less substrate specific and can catalyze deamination of either Phe **26** or Tyr **27**. Detailed studies now need to be undertaken to prove (or disprove) this particular hypothesis.

A considerable amount of research has focused on the deamination step catalyzed by PAL, which is often *speculated* to be a key regulatory step in phenylpropanoid metabolism.[26] PAL can readily be induced by wounding,[20,27,29] infection,[20,27] and UV illumination.[20,27,30] However, its role as a key regulatory step is uncertain, given that various branchpoint pathways are differentially induced, and most branches are far removed from this deamination step (Figure 4). Nevertheless, PAL represents the entry point into phenylpropanoid metabolism, and has been extensively studied in many different plant species. It is a tetramer with a MW range from 240,000 to 330,000 that can be dissociated into subunits of MW ~55,000 to 85,000.[19] In some species, it apparently exists as a single isoform (i.e., in bamboo shoots (*Bambusa oldhami* Munro),[31] strawberry (*Fragaria ananassa*) fruit,[32] and loblolly pine (*Pinus taeda*) developing xylem[33]).

However, caution must be exercised in drawing definite conclusions from data that appear to reveal the presence of only one isoform until several tissue types and stages of development have been examined for a particular species. In other species, distinct isoforms have been reported (e.g., in alfalfa (*Medicago sativa*) suspension cultures,[34] faba bean (*Vicia faba*),[35] bean (*Phaseolus vulgaris*),[36,37] parsley (*Petroselinum crispum* L.),[38,39] rice (*Oryza sativa* L.),[40] and *Arabidopsis thaliana*.[41] The presence of multiple isoforms of PAL in different species awaits clarification with respect to functions: individual isoforms could regulate entry into end-product specific branches, such as lignins, coumarins, etc., as well as even being TAL specific. Although there is differential expression of PAL isoforms in response to stress or developmental cues, distinct roles for specific isoforms have not been defined, even when β-glucuronidase (GUS) reporter gene strategies have been employed using tobacco[42] and *Arabidopsis*.[41] More recent studies on PAL have focused upon determining its subcellular location in potato (*Solanum tuberosum*)[43] and loblolly pine (*Pinus taeda*).[44] Observations were based on indirect immunogold labeling, which, in both cases, revealed the enzyme to be mainly cytosolic (although it was also associated with the membrane fraction in *Pinus taeda*).

B. BIOSYNTHESIS

1. Hydroxylases

Four microsomal cytochrome P-450 dependent monooxygenases have been reported that catalyze formation of the various cinnamic-acid derivatives containing phenolic or catecholic functionalities. All require NAD(P)H (or other co-factors) and molecular oxygen for activity.[21,45] Indeed, it is the introduction of the phenolic moieties by these enzymes that results in the antioxidant properties of the phenylpropanoids that are observed (see Section V). The aromatic hydroxylation enzymes (see Figure 4) include: (1) the monooxygenase, cinnamate-4-hydroxylase (E.C. 1.14.13.11) catalyzes the conversion of cinnamic acid **12** into *p*-coumaric acid **10**,[21] and has been detected in many different plant species and utilizes (*trans*) *E*- rather than (*cis*) *Z*-substrates; (2) a specific *p*-coumarate-3-hydroxylase from mung bean (*Vigna mungo*) seedlings catalyzes caffeic acid **7** formation;[45] (3) phenolase (equivalent to polyphenol oxidase, tyrosinase, and catechol oxidase) also catalyzes caffeic acid **7** formation.[21] Low substrate specificity (reviewed in Gross[21]), and the discovery of the mung bean, *p*-coumarate-3-hydroxylase,[45] requires a clarification of the relative importance of the specific *p*-coumarate-3-hydroxylase; and (4) ferulic acid-5-hydroxylase catalyzes the conversion of ferulic acid **11** into 5-hydroxyferulic acid **28**, has only been detected in poplar (*Populus deltoides x euramericana*).[46]

Other than these studies, little else has been done on the hydroxylases. Limited work has been conducted on localization: cinnamate-4-hydroxylase has been claimed to be associated with membranes (endoplasmic reticulum and chloroplast lamellae); *p*-coumarate-3-hydroxylase exists in a membrane fraction sedimenting in a sucrose gradient between the mitochondria and endoplasmic reticulum;[45] and phenolases have been proposed to be associated with the chloroplast.[47] There is a need to establish the subcellular location of each enzyme unambiguously.

2. *O*-Methyltransferases (OMTs)

The OMTs (E.C. 2.1.1.6) catalyze the methylation of caffeic **7** and 5-hydroxyferulic **28** acids at positions 3 and 5, respectively. Regardless of plant source, they all appear to exist as single enzymes, but with widely varying substrate specificities depending upon the species.[48,49] Although exceptions occur, gymnosperms typically contain monofunctional OMTs catalyzing ferulic acid **11** formation, whereas angiosperms are bifunctional, apparently differentially catalyzing, depending on the species, formation of both ferulic **11** and sinapic **29** acids (Figure 4).[48] Recently, a caffeoyl CoA 3-*O*-methyltransferase was obtained from parsley cell cultures, and the gene encoding the enzyme was cloned. This first example of a gene specific to phenylpropanoid metabolism[50] has been followed by cloning of OMT genes from aspen (*Populus tremuloides*) developing xylem[51] and *Populus deltoides* x *Populus trichocarpa*.[52] Antibodies to OMT have been used to isolate functionally active cDNA from alfalfa (*Medicago sativa*).[53] No definitive studies to establish the subcellular location of these proteins have been carried out to this point.

3. Hydroxycinnamyl CoA Ligase

The CoA ligases (E.C. 6.2.1.12) catalyze the transformation of (hydroxy)cinnamic acids into their corresponding CoA esters via a two-step process involving acyl adenylated intermediates. The resulting enzymatic products are then channeled not only into phenylpropanoid metabolites (lignins, lignans, etc.), but also into phenylpropanoid-acetate (suberins, alkyl ferulates, flavonoids), phenylpropanoid-shikimate (e.g., chlorogenic acid **39**), and other branch-pathway products. Recent work has focused upon characterizing different CoA ligase isoforms from plant species, such as soybean (*Glycine max*),[54] pea (*Pisum sativum*),[55] petunia (*Petunia hybrida*),[56] poplar (*Populus deltoides* x *euroamericana*),[57] and parsley (*Petroselinum crispum*).[58,59] Attempts to ascribe specific metabolic branchpoint pathway roles to individual isoforms (i.e., to lignins, flavonoids, etc.)[56] have been only partially successful. For example, in petunia leaves (*Petunia hybrida*), three CoA ligase isoenzymes were separated and purified: for isoenzyme Ia, only *p*-coumaric **10** and caffeic **7** acids efficiently served as substrates, whereas form Ib catalyzed formation of both *p*-coumaryl **30** and sinapoyl **32** CoA, and isoenzyme II catalyzed *p*-coumaryl **30** and feruloyl **31** CoA synthesis. The authors proposed that isoenzyme Ia was involved in quinic acid (e.g., chlorogenic **39**) biosynthesis, whereas Ib functioned in the pathway leading to sinapyl alcohol **38**, and isoenzyme II acted in the pathway leading to both coniferyl alcohol **37** and the flavonoids. But, in parsley, the two CoA ligases exhibited similar catalytic activities, and, consequently, no specific roles could yet be specified.[58,59] Information about sites of specific gene expression, protein synthesis, and product accumulation will be required in order to define temporal and spatial correlations of enzyme activity with metabolite formation and accumulation. Some progress in this area is now beginning to be made using GUS reporter gene strategies.[18]

4. Cinnamoyl CoA:NADP Oxidoreductase

Cinnamoyl CoA oxidoreductases (E.C. 1.2.1.44) catalyze the conversion of hydroxycinnamate CoA esters **30–32** into the corresponding aldehydes **33–35**, and have been purified from soybean cell suspension cultures (*Glycine max*),[60,61] spruce cambial sap (*Picea abies* L.),[61] and poplar stems (*Populus deltoides* x *euramericana*).[62] Molecular weights range from 36,000 to 38,000, with no subunits being detected. These Type B oxidoreductases apparently require NADPH as co-factor. From the limited studies carried out to date, substrate specificities of these enzymes show analogous trends to those previously noted for *O*-methyltransferases, i.e., the gymnosperm, Norway spruce cinnamoyl CoA:NADP oxidoreductase utilized feruloyl CoA **31** as preferred substrate over *p*-coumaryl CoA **30** and sinapyl CoA **32**, whereas the angiosperm (soybean) efficiently catalyzed reduction of both CoA derivatives. Little else has been done with these enzymes; none of the genes encoding the enzymes have been cloned, and nothing is known about their role(s) in regulation or about their subcellular locations.

5. Cinnamyl Alcohol Dehydrogenase (CAD)

This enzyme (E.C. 1.1.1.195) was first discovered by Mansell et al.[63] in 1974 in extracts of the angiosperm *Forsythia suspensa*. It catalyzes the reduction of aldehydes **33–35** to afford the corresponding monolignols **36–38**. It has been shown, at least in the *Forsythia* sp., to be a Type A reductase of MW ~80,000. In the few examples studied, substrate specificity studies again revealed substantial differences between gymnosperm and angiosperm CAD. Thus, spruce (*Picea abies*) CAD (MW ~72,000)[61] and Japanese black pine (*Pinus thunbergii*) CAD (MW ~67,000)[64] both catalyze the reduction of coniferaldehyde **34** and *p*-coumaraldehyde **33**, but not of sinapaldehyde **35**. In contrast, CAD from angiosperms, such as *Forsythia*, catalyzes the reversible reduction of all of three substrates **33–35**. Recent studies[65,66] have been directed towards cloning the CAD gene from the bean (*Phaseolus vulgaris*). Poplar CAD antibodies were used to immunoprecipitate a putative CAD (MW ~65,000) from fungal elicitor-induced bean cell suspension cultures, and ultimately, a "CAD cDNA clone" was obtained. Sequence data established it to be a malic enzyme and not CAD.[67,68] Since then, others have apparently cloned the CAD gene from *Pinus taeda*.[69] Antibodies to this CAD have also been used in an attempt to determine the subcellular compartmentalization of CAD in *Pinus taeda* cell-suspension cultures. While immunogold labeling *suggested* that CAD was mainly in the plastid of these cells, and also in seedlings in the chloroplast; it remains to be established whether there is a transit peptide sequence in the gene thereby providing further support to its *apparent* subcellular location. Smaller immunoresponses were also noted for the endoplasmic reticulum and mitochondria/cytoplasm, which presumably reflect sites of protein synthesis and transport.[70] CAD has also been purified from tobacco (*Nicotiana tabacum*) in preparation for cloning the gene.[71]

C. LIGNINS

In terms of amount, lignins are the most abundant of the phenylpropanoids. Indeed, as far as naturally occurring organic substances are concerned, only the structural carbohydrate polymer, cellulose, occurs in higher quantity. Lignins are formed principally by polymerization of the monolignols, *p*-coumaryl **36**, coniferyl **37**, and sinapyl **38** alcohols;[17,49,72,73] initially, only the *E*-monolignols **36–38** were thought to be involved in lignification, but the exclusive accumulation of *Z*-monolignols (i.e., *Z*-coniferyl **40** and *Z*-sinapyl **41** alcohols) in *Fagus* bark[74-78] has altered that perception.

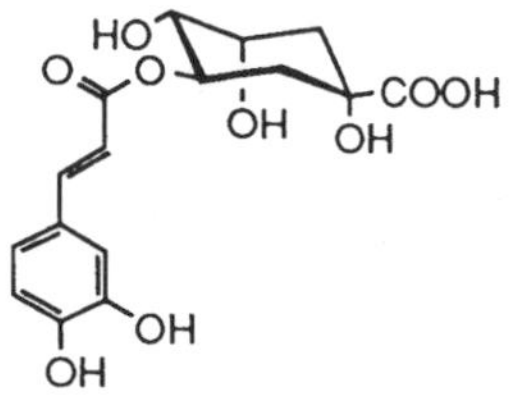

Chlorogenic acid **39**

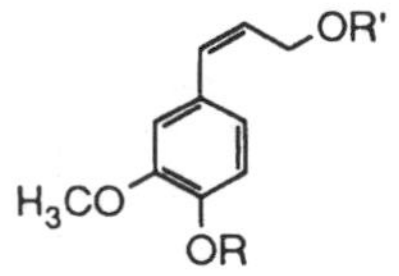

R,R' = H : *Z*-Coniferyl alcohol **40**
R = Glc, R' = H : *Z*-Coniferin **44**
R = H, R' = Glc : *Z*-Isoconiferin **46**

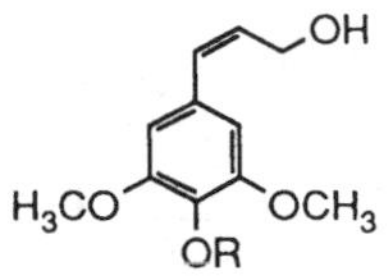

R = H : *Z*-Sinapyl alcohol **41**
R = Glc : *Z*-Syringin **45**

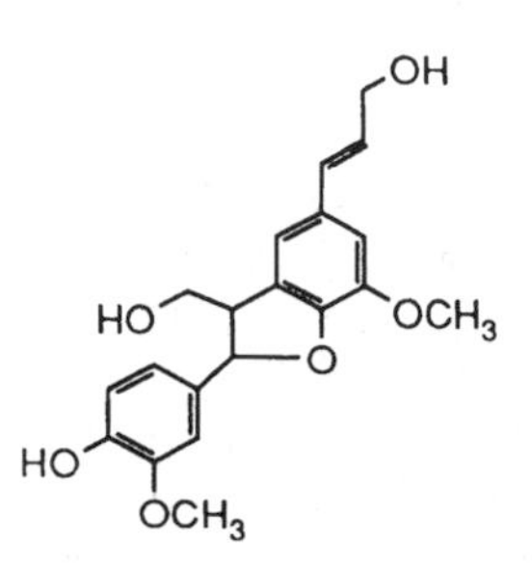

Dehydrodiconiferyl alcohol **47**

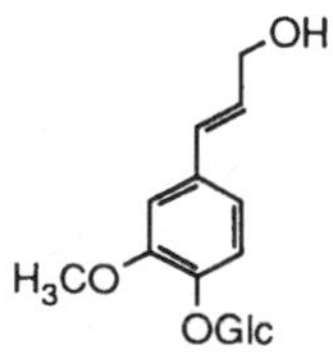

E-Coniferin **42**

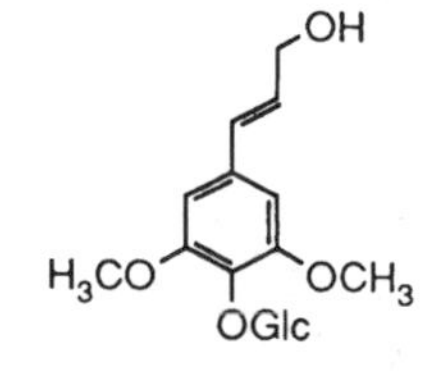

E-Syringin **43**

1. Structure and Relationship to Precursors

The monomeric (i.e., monolignol-derived) composition of lignins varies both within and between species (see Lewis and Yamamoto[17] and Davin and Lewis[49]). Typically, but **not** exclusively, gymnosperm lignins are thought to be derived principally from coniferyl **37** and smaller amounts of *p*-coumaryl **36** alcohols. Woody angiosperms, by contrast, are **largely** derived from both sinapyl **38** and coniferyl **37** alcohols — often in roughly equal proportions — together with small amounts of *p*-coumaryl alcohol **36**. In addition to the monolignols, lignins from grasses and cereals contain hydroxycinnamic acids, such as **10, 11, 28,** and **29** (Figure 4), linked to lignin macromolecules via ester[79] and ether[80] linkages.

Consequently, lignins are viewed as heterogeneous polymers, and are deposited into cell walls during secondary thickening processes — this apparently being initiated at the cell corners and middle lamella, then extending into the secondary wall (see Lewis and Yamamoto[17] and Davin and Lewis[49]). Radiotracer experiments, coupled with microautoradiography, suggest that *p*-coumaryl alcohol **36** is predominantly deposited during early lignification stages, and is then followed by coniferyl alcohol **37**; in angiosperms, a similar trend occurs, except that sinapyl alcohol **38** appears to be deposited during the latter stages of lignin formation.[81] While observations of this type are beginning to provide an explanation for the basis of lignin heterogeneity, they raise fascinating questions as to how this process is regulated and controlled, e.g., (1) how are the individual monolignol moieties temporally and spatially transported across the plasma membrane and into the lignifying cell walls? and (2) what is the relative importance of *E*- and *Z*-monolignols? (It must be emphasized that while much is known about the enzymology of monolignol formation, our understanding of the actual process of lignification is poorly understood.)

Regarding transport, monolignols are frequently found in plant tissue as their *O*-phenolic glucosides, i.e., *E*-coniferin **42**, *E*-syringin **43**, as well as *Z*-coniferin **44**, *Z*-syringin **45**, and *Z*-isoconiferin **46** in *Fagus* sp.[17,49,74-78] It is assumed that the glucosides may represent the chemical species being transferred across the plasma membrane into the lignifying wall; action of cell-wall bound β-glucosidases can then regenerate the monolignols which undergo polymerization (see Yamamoto et al.[82]). The substrate specificity of UDPG:coniferyl alcohol glucosyltransferases is markedly affected by the geometry of the monomer substrate, i.e., in *Fagus* sp., *Z*-coniferyl alcohol **40** was the much preferred substrate when compared to its *E*-analog **37**,[83] whereas in loblolly pine (*Pinus taeda*) both *E*- and *Z*-isomers were efficiently utilized.[84] The ramification of such substrate specificity with respect to lignification awaits a precise clarification. Alternatively, the glucosides may serve in a storage capacity and be used as needed.[82] The precise role of the monolignol glucosides **42–46** will be clarified from the subcellular location of both the UDPG: monolignol glucosyltransferase(s) and their corresponding glucoside **42–46** products.

2. Biosynthesis — Role of Oxidation Chemistry

Products of the phenylpropanoid pathway are exquisite substrates for the free-radical mediated chemistry that constitutes the polymerization pathway for lignin and suberin formation. The enzymology for lignin formation likely involves one electron, H_2O_2-dependent peroxidase-catalyzed transfers that characterize these metalloprotein catalysts and/or O_2-laccase. Free radicals can also be generated by plants in response to stress (e.g., wounding/infection). These may initiate chain reactions that terminate with polymer formation. As these free radicals are encountered, the process may or may not be as controlled or as specific as when polymerization occurs in the normal metabolic sequence described in the following sections.

Following transport, monolignol polymerization is most often described as an H_2O_2-dependent cell-wall bound peroxidase catalyzed reaction (a topic reviewed in detail elsewhere[17,49]). With respect to the putative involvement of specific isoforms, the genes for two strongly anionic peroxidase isoenzymes (pI 3.5 and 3.75) have recently been cloned from tobacco (*Nicotiana tabacum*) and reported to be lignin specific. Although transgenic tobacco[85-87] plants overexpressing or underexpressing these genes have been obtained, a direct correlation with lignification is still lacking; moreover, the transgenic plants overexpressing the strongly anionic peroxidase were stunted and wilted extensively at the onset of flowering, suggesting that processes in addition to lignification were affected.

Laccases (copper-containing monophenol monooxygenases) have also been proposed to function directly in lignification. Polymerization of monolignols *in vitro* forms a lignin-like substance, in a reaction catalyzed by soluble laccases from sycamore (*Acer pseudoplatanus*) suspension cultures.[88] Interestingly, in the earlier literature, peroxidases[89,90] and laccases[91] were both implicated in lignin formation, but the latter lost further consideration[92] when it was reported that a purified laccase from the Japanese lacquer tree, *Rhus vernicifera,* was unable to polymerize coniferyl alcohol **37**. The mechanism of monolignol polymerization has been further clouded by recent reports[93,94] indicating that oxygen-requiring coniferyl alcohol oxidases are responsible for lignification in *Abies balsamea, Larix laricina, Picea rubens, Pinus banksiana,* and *P. strobus*. These oxygen-requiring enzymes apparently catalyze the formation of dehydrodiconiferyl alcohol **47**, pinoresinol **48**, and lignin-like substances from coniferyl alcohol **37**, but it has not yet been established whether they are laccases as could be anticipated. The authors proposed that the oxidases are primarily responsible for lignin formation, but it cannot be excluded that they may also enable peroxidase participation by generating H_2O_2.

These findings bear some similarities to previous investigations by Gross et al.,[95] where it was observed that cell wall suspensions from horseradish (*Armoracia lapathifolia* Gilib) catalyzed the dehydrogenative polymerization of coniferyl alcohol **37** in the absence of exogenously supplied co-factors. This was attributed to the presence of a cell-wall bound enzyme couple — malate dehydrogenase/peroxidase — that converted O_2 into H_2O_2, with the latter then functioning as a cofactor for peroxidase catalyzed polymerization.[96] The rate of peroxidase-induced polymerization was also stimulated by NAD/malate/Mn^{2+} and phenolics.[95,97] A similar situation occurred in *Forsythia* sp.[98] Clearly, much remains to be delineated regarding the biochemistry of monolignol polymerization and its regulation.

Since the structure of lignin provides important clues to the requirements of its synthesis, the study of the *in situ* structure of lignin is relevant. This topic has resisted resolution because there are no suitable methods available to isolate the polymer(s) in chemically unaltered form from plant tissue — hence, all "soluble lignin" preparations are artifacts. As a result, approximations of lignin structure have relied upon the analysis of soluble lignin-derived fragments (see Lewis and Yamamoto[17]) synthetic polymers,[99] — the latter are formed by the dehydrogenative polymerization of monolignols in a reaction catalyzed by peroxidase(s) in the presence of H_2O_2. However, none of these preparations adequately represents lignin structure. Thus, in an effort to overcome these limitations, we pioneered the combined application of carbon-13 specific labeling (of lignin and related phenylpropanoids) and its analysis *in situ,* using solid-state C-13 NMR spectroscopy.[17,49,100-103] This approach has been carried out using lignifying loblolly pine (*Pinus taeda*) cell cultures,[49] as well as *Leuceana leucocephala,*[17,101-103] and wheat (*Triticum aestivum* L.).[100] In this way, the major bonding environments within the polymers have been delineated (see Scheme 1). Importantly, lignin-carbohydrate bonding[17,100-103] was

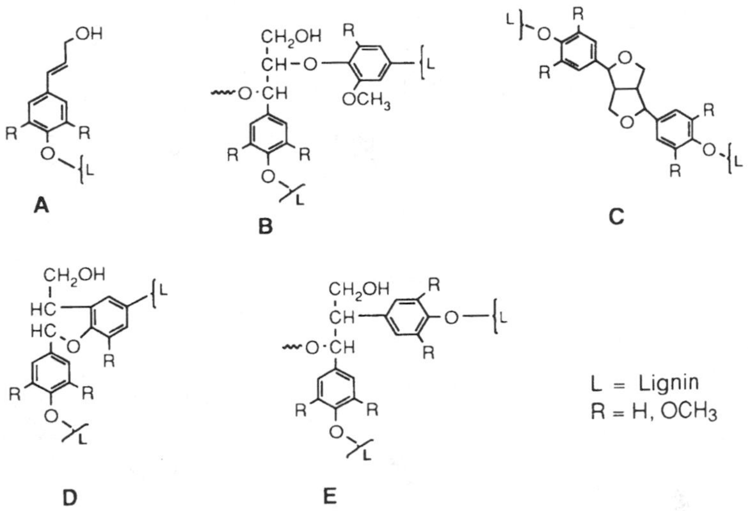

SCHEME 1. Major bonding environments in lignins: L, lignin polymer; R, H or OCH_3.

evident, but more significantly, the frequency of bonding patterns (e.g., 8-*O*-4′) differed substantially from that observed for synthetic dehydrogenatively polymerized lignins. Continued application of this technique is permitting not only the systematic determination of overall lignin structure, but also the investigation of the sequential process of lignification as it relates to cell-wall development and maturation.

D. LIGNANS AND NEOLIGNANS

1. Structures

In addition to the lignin polymers, there is a closely related class of low molecular weight phenylpropanoids, trivially referred to as lignans and neolignans. These metabolites are also directly derived from monolignols (principally coniferyl alcohol **37**), but it is not known how the monolignols are "channeled" into the lignan-neolignan and lignin branches. Lignans and neolignans are most frequently found in plants as "dimers", although higher oligomers also occur.[104] Lignans are typically linked via 8,8′ bonds, whereas neolignans are principally linked via 3,3′, 8,3′, and 8-*O*-4′ bonds (for representative examples, see Scheme 2).[49] At first,[105] only several hundred lignans and neolignans were known, but this number has grown enormously, and they are widely distributed throughout the plant kingdom, ranging from woody plants to constituents of vegetable fiber and grain.[49]

2. Biosynthesis

Until recently, the biosynthesis of this important class of compounds remained obscure. Lignans and neolignans are most often found enantiomerically pure, but the mechanisms by which the achiral precursors undergo stereoselective coupling and their subsequent postcoupling enantioselective transformations have eluded

Neolignans :

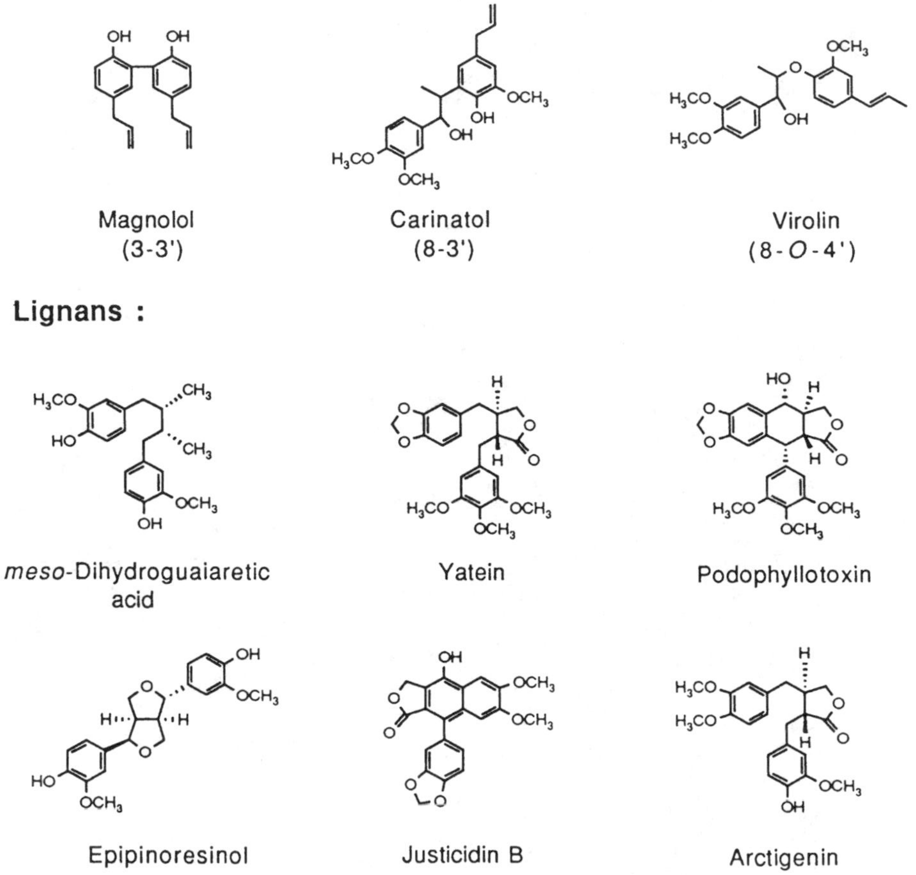

Magnolol (3-3')

Carinatol (8-3')

Virolin (8-*O*-4')

Lignans :

meso-Dihydroguaiaretic acid

Yatein

Podophyllotoxin

Epipinoresinol

Justicidin B

Arctigenin

SCHEME 2. Representative examples of lignans and neolignans.

definition for several decades. The prognosis is excellent, however, to place this apparently chaotic branch of phenylpropanoid metabolism onto a sound scientific footing. This assertion is based on our investigations of lignan biosynthesis in the *Forsythia* sp.[106-110] We have recently discovered that an insoluble enzyme preparation from *Forsythia suspensa* — following removal of readily soluble enzymes — engenders the long sought-after direct stereoselective coupling of two coniferyl alcohol **37** moieties to afford (+)-pinoresinol **48a**; this conversion occurs in the absence of exogenously supplied cofactors, but the rate of (+)-pinoresinol **48a** formation is enhanced significantly[106] under conditions described earlier for H_2O_2-generation.[95] Interestingly, this enzyme preparation displays a strong substrate preference: sinapyl alcohol **38** does not serve as a substrate for stereoselective coupling to afford (+)-syringaresinol **49a**. Thus, it is now important to establish the precise nature of the stereoselective coupling enzyme(s) and to determine its relationship (if any) to lignification.

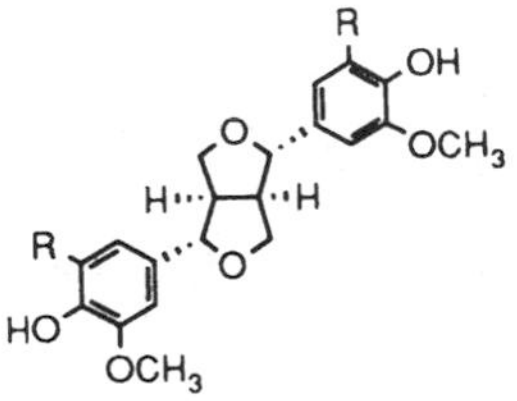

(+)
48a, R = H
49a, R = OCH_3

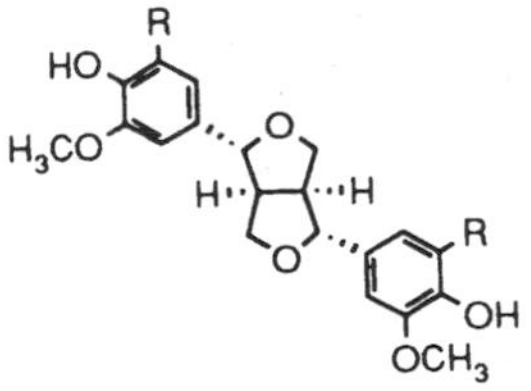

(-)
48b, R = H
49b, R = OCH_3

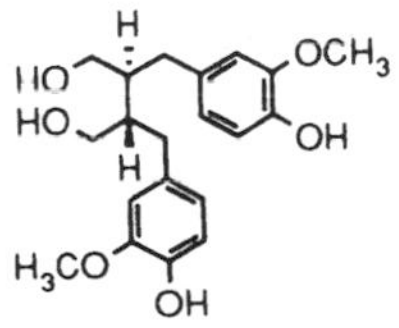

(+)
50a

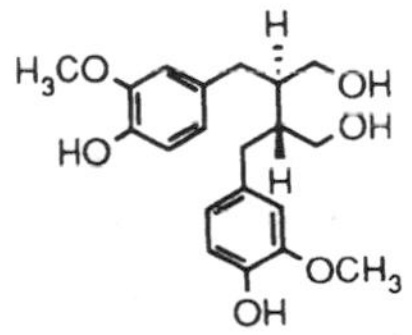

(-)
50b

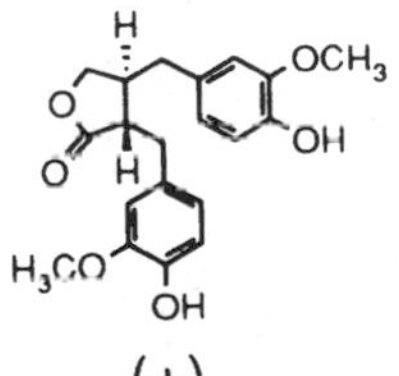

(+)
51a

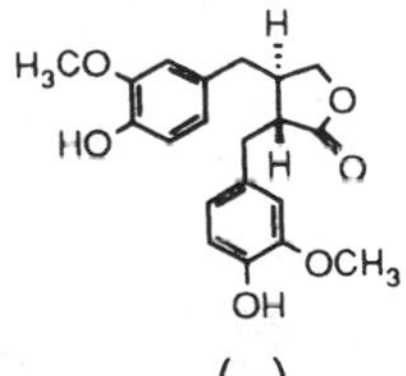

(-)
51b

The formation of other lignans in the *Forsythia* sp. has also been the subject of our investigations: beyond (+)-pinoresinol **48a**, it has been found that (−)-secoisolariciresinol **50b** synthesis occurs via an enantioselective NAD(P)H-dependent reduction of (+)-pinoresinol **48a** to first yield (−)-secoisolariciresinol **50b**,[107-110] which then undergoes an enantiospecific NAD(P)H-dependent dehydrogenation to give (−)-matairesinol **51b**.[108-110] In all cases examined, the corresponding antipodes did not serve as substrates;[108-110] thus, the overall pathway to the *Forsythia* lignans is shown in Figure 5. From the results obtained to date, it should be self-evident that the opportunity to define the enzymology of lignan biogenesis in *Forsythia* sp., as well as establishing tissue specificity of formation, subcellular localization of enzymes and metabolites, and the application of molecular biological techniques is rapidly approaching.

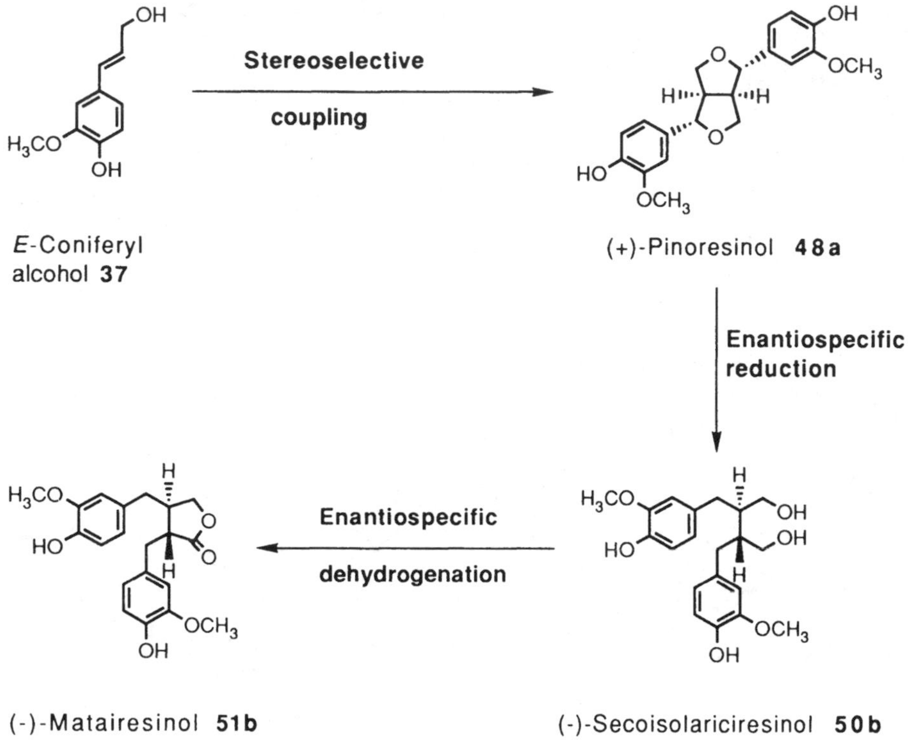

FIGURE 5. Proposed biosynthetic pathways to the lignans, (+)-pinoresinol **48a**, (−)-secoisolariciresinol **50b**, and (−)-matairesinol **51b** in *Forsythia* species.

E. HYDROXYCINNAMIC ACIDS

This subsection is restricted to some of the simplest examples of compounds derived exclusively from the phenylpropanoid pathway. Hydroxycinnamic acids are ubiquitous throughout the plant kingdom, but are most prevalent in cereals, grasses, etc.[17,82] They range in complexity from the monomeric hydroxycinnamic acids, such as compounds **10, 11, 28,** and **29** (see Figure 4), to dimers such as dehydrodiferulic acid **52,**[111] diferuloyl lactone **53,**[112] and 4,4′-dihydroxytruxillic **54** acid, etc.[111] As far as the monomeric hydroxycinnamic acids are concerned, the *E*-derivatives are the initial metabolic products, but both *E*- and *Z*-isomers are formed due to light-induced isomerism during plant growth. Little is known about the subsequent formation of the dimers **52–54**; no purified enzymes have been reported catalyzing the synthesis of dehydrodiferulic acid **52** or diferuloyl lactone **53**, although a specific peroxidase and an oxidase are often assumed to be responsible. Formation of 4,4′-dihydroxytruxillic acid **54** is considered to be due to photochemical dimerization of adjacent *p*-coumaroyl **10** moieties.

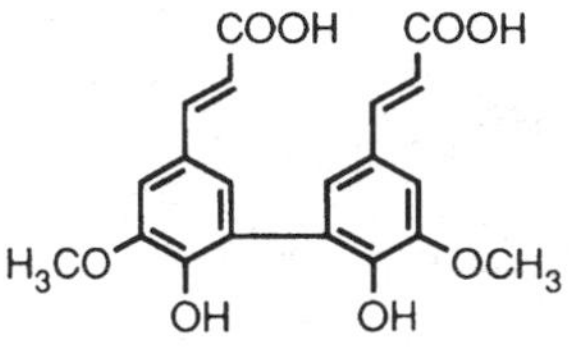

Dehydrodiferulic acid **52**

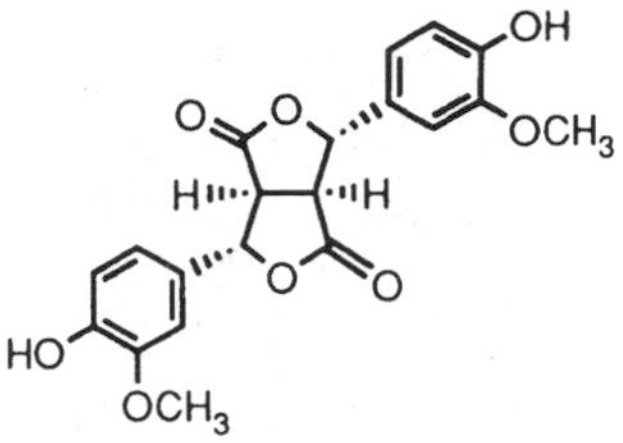

Diferuloyl lactone **53**

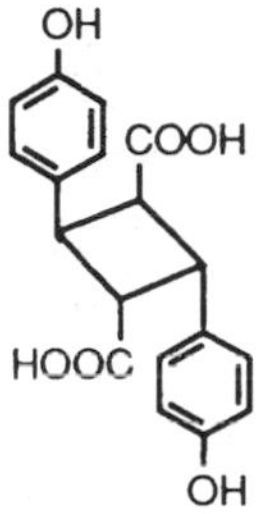

4,4'-Dihydroxytruxillic acid **54**

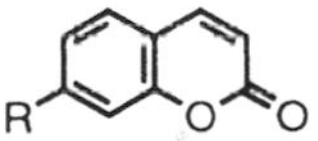

R = H, Coumarin **55**
R = OH, Umbelliferone **56**

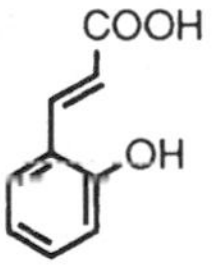

o-Coumaric acid **57**

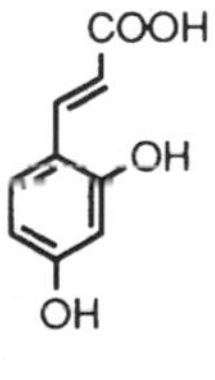

2,4-Dihydroxy-cinnamic acid **58**

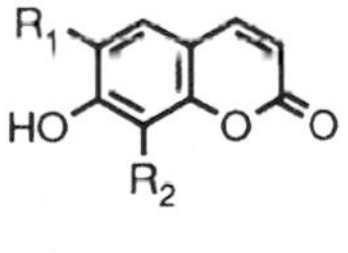

R_1 = OH, R_2 = H Aesculetin **59**
R_1 = H, R_2 = OH Daphnetin **60**

F. COUMARINS

The coumarins[113] contain as their central structural feature the 2H-benzopyran-2-one nucleus; the simplest examples in vascular plants are coumarin **55** and umbelliferone **56**, derived from *E*-cinnamic **12** and *p*-coumaric **10** acids, respectively. Knowledge about the biosynthesis of this class of metabolites is still in disarray. Previous reports using sweet clover (*Melilotus alba*)[114] and petunia (*Petunia hydrida*)[115] indicated that a specific chloroplast-located *ortho*-hydroxylase converted *E*-cinnamic acid **12** into *o*-coumaric acid **57**. However, the *M. alba* findings could not be repeated,[116] and, thus, the evidence for the *o*-hydroxylase is still lacking. Beyond this point, *o*-coumaric **57** and 2,4-dihydroxycinnamic **58** acids can then undergo either light-induced *E-Z* photoisomerization[117] to yield, upon cyclization, the corresponding coumarins **55** or **56**, or glucosylation to afford the *E*-glucosides. If the latter occurs, *E-Z* photoisomerization, as before, yields the *Z*-glucosides, which can undergo spontaneous lactonization to give the corresponding coumarins, following the action of *cis*-specific β-glucosidases. Depending upon the species, additional hydroxylation steps can then occur, e.g., umbelliferone **56** serves as the precursor of aesculetin **59**[118] and daphnetin **60**.[113] None of the enzymes in coumarin biosynthesis has been purified to homogeneity, and their subcellular locations have not been determined.

IV. PHENYLPROPANOID-ACETATE AND OTHER MIXED PATHWAY PHENYLPROPANOID METABOLITES

In addition to the products exclusively derived from the phenylpropanoid pathway, nature is also resplendent with "mixed" pathway metabolites that are only partly of phenylpropanoid origin. The major classes of these phenolics are the alkyl ferulates, suberin, and the flavonoids; these are of phenylpropanoid-acetate pathway origin. Only the compounds related to alkyl ferulates and suberin formation will be discussed.

A. ALKYL FERULATES

Some of the simplest examples are long-chain alkyl hydroxycinnamate esters, such as the feruloyl fatty acid esters **61–69** that accumulate in the periderm of potato (*Solanum tuberosum*) tubers;[119] phellochryseine **70** (isolated as its methyl ester) from cork;[120] the even-chain length C_{16}-C_{28} alkyl ferulate esters **61, 62, 64, 66, 68,** and **69** and the even-chain length esters from Douglas fir (*Pseudotsuga menziesii*) bark.[121,122] The presence of these long-chain alkyl ferulates, as well as related hydroxyphenyl alkanoic acid esters, in the periderm of a variety of plants suggests an important role for them, or their derivatives, in suberization. However, this has not been established.

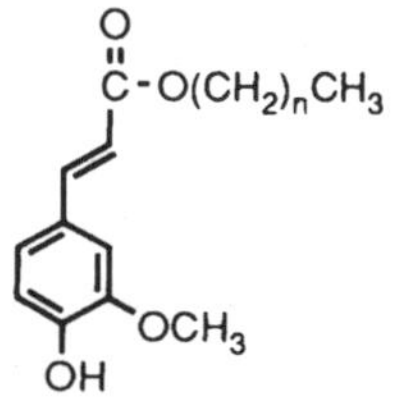

$C(=O)$-$O(CH_2)_nCH_3$; OCH_3 ; OH

61-69,
n = 15,17,18-21,
23, 25 and 27

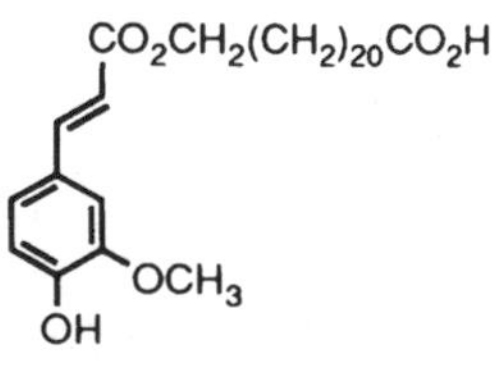

$CO_2CH_2(CH_2)_{20}CO_2H$; OCH_3 ; OH

Phellochryseine **70**

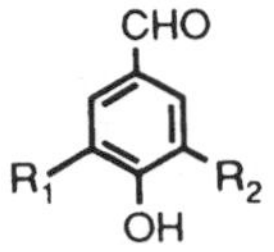

CHO / OH / R_1 / R_2

71, R_1, R_2 = H
72, R_1 = H, R_2 = OCH_3
73, R_1, R_2 = OCH_3

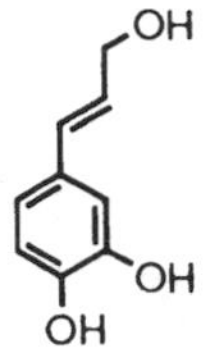

OH / OH / OH

Caffeoyl
alcohol **74**

The biochemistry of alkyl ferulate formation (and related analogues) has not been elucidated, in part because of difficulties in separating the various mixtures of alkyl ferulates in plants. Consequently, in a study directed towards determining the biochemistry of both alkyl ferulate and suberin formation, we developed a rapid, reliable high-performance liquid chromatographic separation of the alkyl ferulates in potato periderm.[119] A time course of their accumulation in wound-healing potato tubers revealed that they were at or near nondetectable levels until 3 to 5 days after wounding, when they could be readily detected over the next 9 days. Mechanical separation and analysis of the wound periderm from 7-day wound-healed tuber slices showed that the ferulates were restricted to this layer, and were not detected in underlying tissue layers. Thus, alkyl ferulate accumulation was temporarily and spatially coincident to the deposition of a suberized diffusion barrier that appeared after 4 to 6 days in wound-healing tuber disks. Ferulates did not accumulate to high levels until after an effective suberin layer was formed. Possibly ferulates, synthesized early in the wound-healing process, were covalently incorporated into the suberin polymer. Consequently, it needs to be established if ferulates serve directly as suberin precursors, if they first undergo covalent modification of the side-chain prior to suberin formation, or if other substrates are involved.

B. SUBERINS

Essentially, all that is known about the biosynthesis of the aromatic domain of suberin is that phenylalanine ammonia-lyase and peroxidase activities are wound inducible,[123] and that this induction correlates with suberization. Their induction, however, offers little insight into the biosynthetic pathways that lead to the aromatic monomers of suberin. Interestingly, a peroxidase deemed to be "suberin-specific" has been purified,[124] and the gene cloned.[125,126] The substrate specificity of this enzyme has not yet been deduced, nor has it yet been proven to be involved in the "final assembly" of either aromatic or aliphatic domains of suberin.

1. Structure

The term was apparently first coined by Chevreul in 1815 to describe a primary constituent of bottle cork insoluble in alcohol or water. It is a polymeric substance, considered to contain both aliphatic and aromatic domains, laid down in a lamellar-like fashion within the cell walls of certain vascular plant tissues.[127-128] The structure of suberin remains essentially undefined in spite of extensive studies that have

Major Components	Example	Formula
1-Alkanols	1-Eicosanol	$HO\text{-}(CH_2)_{19}\text{-}CH_3$
Alkanoic acids	Docosanoic acid	$HO_2C\text{-}(CH_2)_{20}\text{-}CH_3$
ω-Hydroxyalkanoic acids	22-Hydroxydocosanoic acid	$HO_2C\text{-}(CH_2)_{21}\text{-}OH$
α,ω -Alkanedioic acids	Phellogenic acid	$HO_2C\text{-}(CH_2)_{20}\text{-}CO_2H$

Minor Components	Example	Formula
Hydroxylated-α,ω-alkanedioic acids	Dihydroxyoctadecane-1,18-dioic acid	$HO_2C\text{-}(CH_2)_7(CHOH)_2(CH_2)_7\text{-}CO_2H$
Hydroxylated-ω-hydroxy-alkanoic acids	9,10,18-Trihydroxy-octadecanoic acid	$HOCH_2\text{-}(CH_2)_7(CHOH)_2(CH_2)_7\text{-}CO_2H$

FIGURE 6. Representative aliphatic components of suberin.

delineated the major aliphatic building blocks.[128,129] However, determination of its aliphatic nature can be tracked back to the isolation of suberic acid by La Grange in 1797, following nitric-acid treatment of cork. It took almost another 160 years before the existence of a phenolic component in suberin became apparent.[130] Suberized lamellae[127] are deposited in a number of tissues — principally in periderm of both nonwoody[131] (e.g., sweet potato, etc.) and woody (e.g., oak[132] and Douglas fir[133]) plants, where the suberin polymer can reach concentrations as high as 40 to 50% of the suberized tissue. Suberized lamellae[128] have also been found in the epidermis and hypodermis of roots, e.g., corn, the endodermis (Casparian band), bundle sheaths of grasses, idioblast sheaths between secretory organs (e.g., of glands, trichomes, etc.), pigment strands of grains, and connections between vascular tissue and seed coats. These lamellae are laid down between the primary cell wall and the plasmalemma,[128] hence, implicating a requirement for extracellular substrates and enzymes. As for the lignins, no method exists to isolate suberin in either a pure form or in its native state.

Much of our limited knowledge about suberin structure comes largely from chemical degradation studies in which the components released from suberin-enriched tissues were isolated and subsequently characterized using GC-MS and $^1H/^{13}C$ NMR spectroscopic techniques.[134-136,142] It was initially established that long-chain fatty acids, from C_{16} to $>C_{28}$ alcohols, and ω-hydroxyalkanoic and α,ω-alkanedioic acids predominated (Figure 6).[137-142] Rare, odd-chain length analogs have also been reported in the hydrolysates of suberin-enriched samples.[143] Findings of this type led to the initial conclusion that suberin was a readily saponifiable ester-linked lipid-like material (i.e., a polyester).[144,145] Suberized tissue also contained phenolic material, since sinapic acid **29** was isolated from saponified suberin samples, where it was presumed linked to lipid material.[130]

Further evidence[146] for involvement of hydroxycinnamic acids in suberin was established when potato and beet "suberin-enriched" tissues were treated with $LiAl^2H_4$; among the products formed was trideuterated (presumably $8,9-^2H_3$) con-

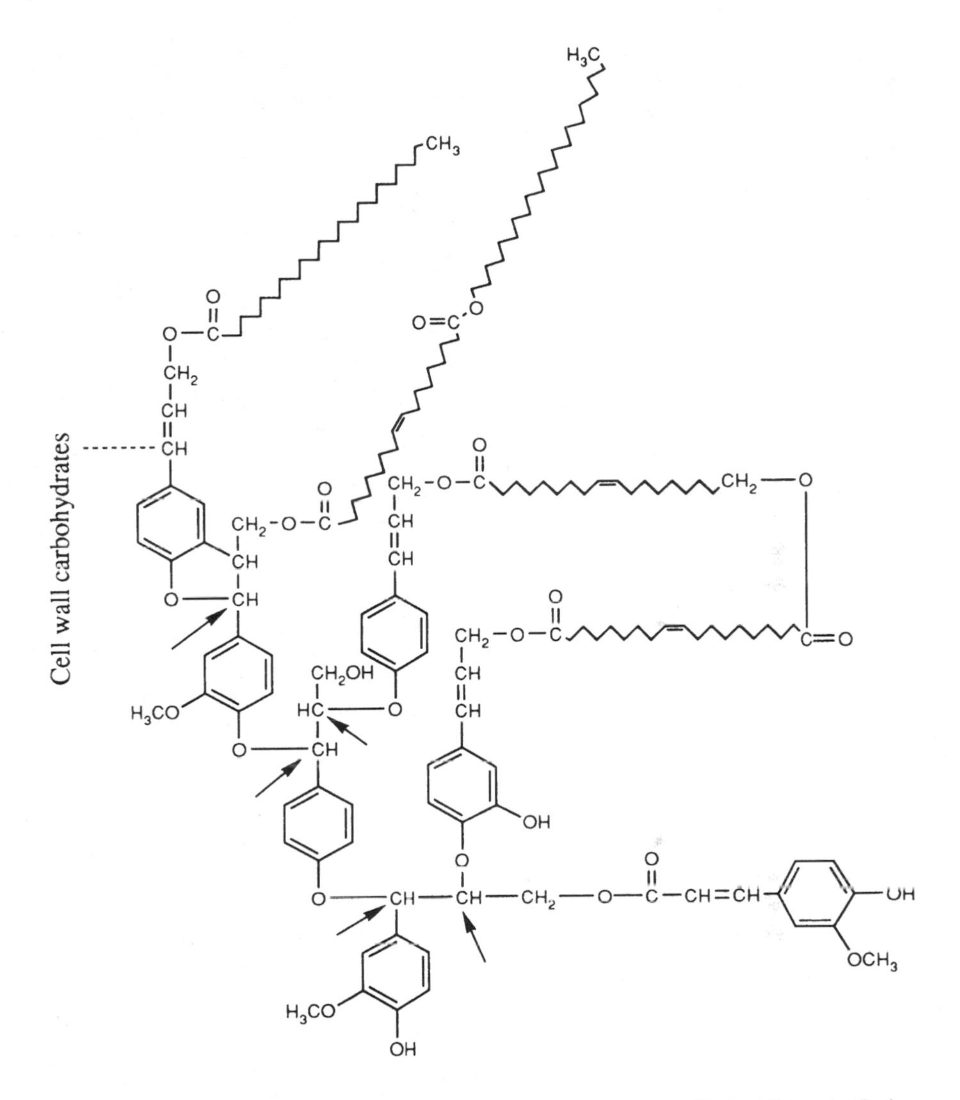

FIGURE 7. Proposed tentative model for the structure of suberin (→ = alkylaryl linkage). (Redrawn from Kolattukudy, P. E., *Can J. Bot.*, 62, 2918, 1984. With permission.)

iferyl alcohol **37**, which could only result via feruloyl ester reduction. Hydroxy-cinnamic acids and hydroxycinnamides have also been isolated from cell walls of stressed potato tubers, but no direct role in suberization has been established.[147]

A tentative "lignin-like" model for suberin was next proposed[128] where both aromatic and aliphatic components were covalently linked (Figure 7). Note that the phenolic domains are proposed derived from the monolignols **36** and **37** (and caffeoyl alcohol **74**), as well as from ferulic acid **11** linked together via alkylaryl ether bonds (designated →). On the other hand, the aliphatic domains are envisaged to consist largely of alkali-sensitive long-chain (C_{18} to C_{28}) fatty acids or alcohols esterified to each other and to the aromatic domain. To probe further the phenolic composition of suberin, the following experiments were carried out and the results obtained were as follows:

1. The insoluble fraction after extraction of periderm tissue from wound-healing potato tubers, was subjected to nitrobenzene oxidation;[148] and it yielded *p*-hydroxybenzaldehyde **71** and vanillin **72** together with traces of syringaldehyde **73** in agreement with the proposed model.
2. Separate administration of [U – ^{14}C]phenylalanine **26** and [U – ^{14}C]cinnamic acid **12** to wound-healing potato tissue, followed by nitrobenzene oxidation resulted in radiolabeled *p*-hydroxybenzaldehyde **71** and vanillin **72**.
3. When Borg-Olivier and Monties[149] applied their recently developed thioacidolysis procedure to characterize suberin in suberized potato periderm tissue, only guaiacyl (G) and syringyl (S) derivatives were released in small amounts, in a 2:1 ratio; the corresponding *p*-hydroxyphenylpropane derivatives were not detected.
4. The aromatic component of suberin-enriched preparations has also been examined, using both ^{1}H and ^{13}C-NMR spectroscopy.[134-136] Pretreatment of suberized tissue from potato-wound periderm with cellulases and pectinases, followed by solvent extraction and treatment of the preparation with BF_3-methanol, yielded a suberin-rich preparation, the ^{13}C-NMR spectrum of which showed strong carboxylic acid and methylenic signals, as well as aromatic signals. This preparation showed greatly reduced carbohydrate signals compared to samples not treated with the cell-wall degrading enzymes, with no apparent changes in the resonances attributed to suberin. (Application of specific ^{13}C labeling studies would presumably provide a means to determine the interunit linkages present in suberin, as has been accomplished for lignins.[99-103])
5. Finally, in another study, a suberin-derived preparation was released from suberized potato tissue by adaptation of the dioxane-H_2O method used to solubilize lignin from plant tissue.[136] This preparation lacked any "typical" lignin bonding patterns (e.g., alkylaryl ethers) in the aromatic component of the polymer as evidenced by ^{1}H and ^{13}C-NMR spectroscopic techniques, thereby, once again raising doubts over the proposed suberin structure in Figure 7.

Taking these data into account, combined with that from saponification, thioacidolysis, alkaline nitrobenzene oxidation, and $LiAlH_4$ reduction, it is clear that the chemical nature of the aromatic component of suberin and its interunit bonding patterns are not consistent with a typical "lignin-like" molecule, and it requires redefinition.

2. Biochemistry of Suberization

Most enzymological studies have been directed toward understanding the formation of the aliphatic domain of suberin, and focused upon the biosynthesis of the predominant components, ω-hydroxy and alkanedioic acids (see Kolattukudy[128] for review). Three enzymes have been described: (1) an ω-hydroxylase which catalyzes the formation of ω-hydroxyalkanoic acids; (2) a constitutive ω-hydroxy alkanoic acid dehydrogenase which forms the corresponding ω-oxoalkanoic acid; and (3) a wound-inducible ω-oxoalkanoic acid dehydrogenase that oxidizes the oxoderivative to the corresponding dicarboxylic acid. These enzymes were postulated to generate the various α,ω-alkanedioic acids characteristic of suberin, but, surprisingly, only the commercially available ω-hydroxyhexadecanoic acid was evaluated as a substrate. Importantly, compounds such as alkyl ferulates were not tested as substrates, even though the oxidation of the terminal methyl of the alkyl chain of these esters would yield compounds such as phellochryseine **70**.

In summary, the following is unknown with respect to the biosynthesis of the phenolic domain of suberin: (1) the chemical identity of the phenylpropanoid moiety (or moieties) undergoing transport from the cytosol into the suberizing wall and how this transport is affected by stress; (2) if fatty acid/fatty alcohol conjugation with phenolic acids/alcohols occurs prior, or subsequent, to deposition in the wall, or at all; and (3) whether alkyl chains of ferulate esters are modified (i.e., terminal hydroxylation and/or carboxylation) prior to suberization, or, instead, whether different acyl moieties are conjugated; (4) the role of enzymes, e.g., in generating H_2O_2 and the requirements for free-radical chemistry in suberin formation.

V. SUMMARY

Collectively, secondary plant metabolites represent a complex, diversity of organic products. The emphasis of this chapter precludes discussion of the varied important functions served by flavonoids and their derivatives. But a common thread of understanding that ties together the phenylpropanoid-acetate pathway metabolites (such as the suberins, oligomeric flavonoids, e.g., proanthocyanidins or condensed tannins) is their susceptibility to phenylpropanoid and oxidative, free-radical chemistry. The formation of insoluble polymers as end-products of this chemistry provides the normal, protective barriers from the environment under unstressed conditions, and also establishes the boundaries that separate normal from damaged tissue. Among the stress responses that lead to this chemistry are those that cause oxidative damage. As noted throughout the text, the definition of a subcellular location for enzymes and substrates is an essential step in determining the specificity of responses to individual stresses. Since this metabolism functions to protect plant cells, future strategies to develop plants with improved tolerance to stress may require its manipulation. The demanding work of establishing polymer structures formed under normal and stressed conditions will yield important information about the control and regulation of the polymerization events. Detailed requirements for enzyme isoforms and substrates will provide information upon which to base selection of phenotypes and to develop rational genetic engineering strategies for improving a plant's capacity to accommodate oxidative stress.

Much biochemistry of phenylpropanoids and the related phenylpropanoid-acetate pathway products remains to be described. Increasingly sophisticated analytical techniques provide improved opportunities to define accurate structures and reactant/product relationships. Protein sequence analyses promise the rapid integration of new information from plants into the more comprehensive literature on animal systems. These approaches, along with the application of the tools of molecular biology, will establish a more complete understanding of how oxidative chemistry functions in the metabolism of phenols and of how cellular compartmentation contributes to its regulation.

ACKNOWLEDGMENTS

The author wishes to thank the U. S. Department of Energy (DE-FG-06-91ER20022) for financial assistance.

REFERENCES

1. **Haddock, E. A., Gupta, R. K., Al-Shafi, S. M. K., Layden, K., Haslam, E., and Magnolato, D.,** The metabolism of gallic acid and hexahydroxydiphenic acid in plants: biogenetic and molecular taxonomic considerations, *Phytochemistry,* 21, 1049, 1982.
2. **Saijo, R., Nonaka, G. I., and Nishioka, I.,** Gallic acid esters of bergenin and norbergenin from *Mallotus japonicus, Phytochemistry,* 29, 267, 1990.
3. **Yoshida, T., Seno, K., Takama, Y., and Okuda, T.,** Bergenin derivatives from *Mallotus japonicus, Phytochemistry,* 21, 1180, 1982.
4. **Lewis, N. G. and Yamamoto, E.,** Tannins: their place in metabolism, in *Chemistry and Significance of Condensed Tannins,* Hemingway, R. W. and Karchesy, J. J., Eds., Plenum Press, New York, 1989, 23.
5. **Amrhein, N., Topp, H., and Joop, O.,** The pathway of gallic acid biosynthesis in higher plants, *Plant Physiol. Suppl.,* 75, 18, 1984.
6. **Becerril, J. M., Duke, S. O., and Lydon, J.,** Glyphosate effects on shikimate pathway products in leaves and flowers of velvetleaf, *Phytochemistry,* 28, 695, 1989.
7. **Saijo, R.,** Pathway of gallic acid biosynthesis and its esterification with catechins in young tea shoots, *Agric. Biol. Chem.,* 47, 455, 1983.
8. **Zenk, M. H.,** Zur frage der biosynthese von gallussaure, *Z. Naturforsch.,* 196, 83, 1964.
9. **El-Basyouni, S. A., Chen, D., Ibrahim, R. K., Neish, A. C., and Towers, G. H. N.,** The biosynthesis of hydroxybenzoic acids in higher plants, *Phytochemistry,* 3, 485, 1964.
10. **Yazaki, K., Heide, L., and Tabata, M.,** Formation of *p*-hydroxybenzoic acid from *p*-coumaric acid by cell-free extract of *Lithospermum erythrorhizon* cell cultures, *Phytochemistry,* 30, 2233, 1991.
11. **McMillan, C.,** The condensed tannins (proanthocyanidins) in seagrass, *Aq. Bot.,* 20, 351, 1984.
12. **Peng, S., Scalbert, A., and Monties, B.,** Insoluble ellagitannins in *Castanea sativa* and *Quercus petraea* woods, *Phytochemistry,* 30, 775, 1991.
13. **Gross, G. G.,** Enzymology of gallotannin biosynthesis, in *Plant Cell Wall Polymers: Biogenesis and Biodegradation,* Lewis, N. G. and Paice, M. G., Eds., *ACS Symp. Ser.,* 399, 108, 1989.
14. **Denzel, K. and Gross, G. G.,** Biosynthesis of gallotannins: enzymatic "disproportionation" of 1,6 digalloylglucose to 1,2,6 trigalloylglucose and 6-galloylglucose by an acyltransferase from leaves from *Rhus typhina* L., *Planta,* 184, 285, 1991.

15. **Hoffman, A. S. and Gross, G. G.,** Biosynthesis of gallotannins: formation of polygalloylglucose by enzymatic acylation of 1,2,3,4,6-penta-*O*-galloylglucose, *Arch. Biochem. Biophys.*, 283, 530, 1990.
16. **Hillis, W. E.,** Biosynthesis of tannins, in *Biosynthesis and Biodegradation of Wood Components*, Higuchi, T., Ed., Academic Press, New York, 1985, 325.
17. **Lewis, N. G. and Yamamoto, E.,** Lignins: occurrence, biosynthesis and biodegradation, *Annu. Rev. Plant Physiol. and Plant Mol. Biol.*, 41, 455, 1990.
18. **Hauffe, K. D., Paszowski, U., Schulze-Lefert, P., Hahlbrock, K., Dangl, J. L., and Douglas, C. J.,** A parsley 4 CL-1 promoter fragment specifies complex expression patterns in transgenic tobacco, *Plant Cell*, 3, 4350, 1991.
19. **Hanson, K. R. and Havir, E. A.,** Phenylalanine ammonia-lyase, in *The Biochemistry of Plants*, Vol. 7, Conn, E. E., Ed., Academic Press, New York, 1981, 577.
20. **Jones, D. H.,** Phenylalanine ammonia-lyase: regulation of its induction and its role in plant development, *Phytochemistry*, 23, 1349, 1984.
21. **Gross, G. G.,** Biosynthesis and metabolism of phenolic acids and monolignols, in *Biosynthesis and Biodegradation of Wood Components*, Higuchi, T., Ed., Academic Press, Orlando, FL, 1985, 229.
22. **Hanson, K. R., Wightman, R. H., Staunton, J., and Battersby, A. R.,** *J. Chem. Soc. Chem. Commun.*, 185, 1971.
23. **Ife, R. and Haslam, E.,** The shikimate pathway. Part III. Stereochemical course of the L-phenylalanine ammonia lyase reaction, *J. Chem. Soc. (C)*, 16, 2818, 1971.
24. **Neish, A. C.,** Formation of *m*- and *p*-coumaric acids by enzymatic deamination of the corresponding isomers of tyrosine, *Photochemistry*, 1, 1, 1961.
25. **Ellis, B. W., Zenk, M. H., Kirby, G. W., Michael, J., and Floss, H. G.,** Steric course of the tyrosine ammonia lyase reaction, *Phytochemistry*, 12, 1057, 1973.
26. **Liang, X., Dron, M., Cramer, C. L., Dixon, R. A., and Lamb, C. J.,** Differential regulation of phenylalanine ammonia-lyase genes during plant development and by environmental cues, *J. Biol. Chem.*, 264, 14486, 1989.
27. **Hahlbrock, K. and Scheel, D.,** Physiology and molecular biology of phenylpropanoid metabolism, *Annu. Rev. Plant Physiol. Plant Mol. Biol.*, 40, 347, 1989.
28. **Tanaka, Y. and Uritani, I.,** Purification and properties of phenylalanine ammonia-lyase in cut-injured sweet potato, *J. Biochem.*, 81, 963, 1977.
29. **Tanaka, Y., Matsuoka, M., Yamamoto, N., Ohashi, Y., Kano-Murakami, Y., and Ozeki, Y.,** Structure and characterization of a cDNA clone for phenylalanine ammonia-lyase from cut-injured roots of sweet potato, *Plant Physiol.*, 90, 1403, 1989.
30. **Havir, E. A. and Hanson, K. R.,** L-Phenylalanine ammonia-lyase. I. Purification and molecular size of the enzyme from potato tubers, *Biochemistry*, 7, 1896, 1968.
31. **Chen, R.-Y., Chang, T.-C., and Liu, M.-S.,** Phenylalanine ammonia-lyase of bamboo shoots, *Agric. Biol. Chem.*, 52, 2137, 1988.
32. **Given, N. K., Venis, M. A., and Grierson, D.,** Purification and properties of phenylalanine ammonia-lyase from strawberry fruit and its synthesis during ripening, *J. Plant Physiol.*, 133, 31, 1988.
33. **Whetten, R. W. and Sederoff, R. R.,** Phenylalanine ammonia-lyase from loblolly pine, *Plant Physiol.*, 98, 380, 1992.
34. **Jorrin, J. and Dixon, R. A.,** Stress responses in alfalfa (*Medicago sativa* L.). II. Purification, characterization, and induction of phenylalanine ammonia-lyase isoforms from elicitor-treated cell suspension cultures, *Plant Physiol.*, 92, 447, 1990.
35. **Lopez-Valbuena, R., Obrero, R., Jorrin, J., and Tena, M.,** Isozyme multiplicity in phenylalanine ammonia-lyase from *Vicia faba* leaves, *Plant Physiol. Biochem.*, 29, 159, 1991.
36. **Bolwell, G. P., Bell, J. N., Cramer, C. L., Schuch, W., Lamb, C. J., and Dixon, R. A.,** Phenylalanine ammonia-lyase from *Phaseolus vulgaris*; characterization and differential induction of multiple forms from elicitor-treated cell suspension cultures, *Eur. J. Biochem.*, 149, 411, 1985.

37. **Cramer, C. L., Edwards, K., Dron, M., Liang, X., Dildine, S. L., Bolwell, G. P., Dixon, R. A., Lamb, C. J., and Schuch, W.,** Phenylalanine ammonia-lyase gene organization and structure, *Plant Mol. Biol.*, 12, 367, 1989.
38. **Schulz, W., Eiben, H.-G., and Hahlbrock, K.,** Expression in *Escherichia coli* of catalytically active phenylalanine ammonia-lyase from parsley, *FEBS Lett.*, 258, 335, 1989.
39. **Lois, R., Dietrich, A., Hahlbrock, K., and Schulz, W.,** Phenylalanine ammonia-lyase gene from parsley: structure, regulation and identification of elicitor and light responsive *cis*-acting elements, *EMBO J.*, 8, 1641, 1989.
40. **Minami, E., Ozeki, Y., Matsuoka, M., Koizuka, N., and Tanaka, Y.,** Structure and some characterization of the gene for phenylalanine ammonia-lyase from rice plants, *Eur. J. Biochem.*, 185, 19, 1989.
41. **Ohl, S., Hedrick, S. A., Chory, J., and Lamb, C. J.,** Functional properties of a phenylalanine ammonia-lyase promoter from *Arabidopsis, Plant Cell,* 2, 837, 1990.
42. **Liang, X., Dron, M., Schmid, J., Dixon, R. A., and Lamb, C. J.,** Developmental and environmental regulation of a phenylalanine ammonia-lyase-β-glucuronidase gene fusion in transgenic tobacco plants, *Proc. Natl. Acad. Sci. U.S.A.*, 86, 9284, 1989.
43. **Shaw, N. M., Bolwell, G. P., and Smith, C.,** Wound-induced phenylalanine ammonia-lyase in potato (*Solanum tuberosum*) tuber discs. Significance of glycosylation and immunolocalization of enzyme subunits, *Biochem. J.*, 267, 163, 1990.
44. **He, L., Davin, L. B., Sederoff, R. R., and Lewis, N. G.,** Immunogold localization of phenylalanine ammonia lyase in the stem and cell-suspension culture of loblolly pine *(Pinus taeda),* submitted.
45. **Kojima, M. and Takeuchi, W.,** Detection and characterization of *p*-coumaric acid hydroxylase in mung bean (*Vigna mungo*) seedlings, *J. Biochem.*, 105, 265, 1989.
46. **Grand, C.,** Ferulic acid 5-hydroxylase: a new cytochrome P-450-dependent enzyme from higher plant microsomes involved in lignin synthesis, *FEBS Lett.*, 169, 7, 1984.
47. **Sato, M.,** Metabolism of phenolic substances by the chloroplasts. III. Phenolase as an enzyme concerning the formation of esculetin, *Phytochemistry,* 6, 1363, 1967.
48. **Kuroda, H.,** Comparative studies on *O*-methyltransferases involved in lignin biosynthesis, *Wood Res.*, 69, 91, 1983.
49. **Davin, L. B. and Lewis, N. G.,** Phenylpropanoid metabolism: biosynthesis of monolignols, lignans and neolignans, lignins and suberins, in *Recent Advances in Phytochemistry,* Stafford, H. A., Ed., 1992, 325.
50. **Schmitt, D., Pakusch, A. E., and Mattern, U.,** Molecular cloning, induction and taxonomic distribution of caffeoyl-CoA-3-*O*-methyltransferase, an enzyme involved in disease resistance, *J. Biol. Chem.*, 266, 17416, 1991.
51. **Bugos, R. C., Chiang, V. L. C. and Campbell, W. H.,** Seasonal expression of a lignin specific *O*-methyltransferase in aspen wood, *Plant Physiol.*, 96, 84, 1991.
52. **Dumas, B., Van Doorsselaere, J. V., Gielen, J., Legrand, M., Fritig, B., Van Montagu, M., and Inze, D.,** Nucleotide sequence of a complementary DNA encoding *O*-methyltransferase from poplar, *Plant Physiol.*, 98, 796, 1992.
53. **Gowri, G., Bugos, R. C., Campbell, W. H., Maxwell, C. A., and Dixon, R. A.,** Stress responses in alfalfa (*Medicago sativa* L.). X. Molecular cloning and expression of *S*-adenosyl-L-methionine: caffeic acid 3-*O*-methyltransferase, a key enzyme of lignin biosynthesis, *Plant Physiol.*, 97, 7, 1991.
54. **Knobloch, K.-H. and Hahlbrock, K.,** Isoenzymes of *p*-coumarate: CoA ligase from cell suspension culture of *Glycine max, Eur. J. Biochem.*, 52, 311, 1975.
55. **Wallis, P. J. and Rhodes, M. J. C.,** Multiple forms of hydroxycinnamate: CoA ligase in etiolated pea seedlings, *Phytochemistry,* 16, 1891, 1977.
56. **Ranjeva, R., Boudet, A. M., and Faggion, R.,** Phenolic metabolism in petunia tissues. IV. Properties of *p*-coumarate: coenzyme A ligase isoenzymes, *Biochimie,* 58, 1255, 1976.
57. **Grand, C., Boudet, A., and Boudet, A. M.,** Isoenzymes of hydroxycinnamate:CoA ligase from poplar stems. Properties and tissue distribution, *Planta,* 158, 225, 1983.

58. **Lozoya, E., Hoffman, H., Douglas, C., Schulz, W., Scheel, D., and Hahlbrock, K.,** Primary structures and catalytic properties of isoenzymes encoded by the two 4-coumarate:CoA ligase genes in parsley, *Eur. J. Biochem.*, 176, 661, 1988.
59. **Douglas, C., Hoffman, H., Schulz, W, and Hahlbrock, K.,** Structure and elicitor or a U.V.-light stimulated expression of two 4-coumarate:CoA ligase genes in parsley, *EMBO J.*, 6, 1189, 1987.
60. **Wengenmayer, H., Ebel, J., and Grisebach, H.,** Enzymic synthesis of lignin precursors. Purification and properties of a cinnamoyl-CoA:NADPH reductase from cell suspension cultures of soybean *(Glycine max), Eur. J. Biochem.*, 65, 529, 1976.
61. **Luderitz, T. and Grisebach, H.,** Enzymic synthesis of lignin precursors. Comparison of cinnamoyl-CoA reductase and cinnamyl alcohol: $NADP^+$ dehydrogenase from spruce (*Picea abies* L.) and soybean (*Glycine max* L.), *Eur. J. Biochem.*, 119, 115, 1981.
62. **Sarni, F., Grand, C., and Boudet, A. M.,** Purification and properties of cinnamoyl-CoA reductase and cinnamyl alcohol dehydrogenase from poplar stems (*Populus* x *euamericana*), *Eur. J. Biochem.*, 139, 259, 1984.
63. **Mansell, R. L., Gross, G. G., Stöckigt, J., Franke, H., and Zenk, M. H.,** Purification and properties of cinnamyl alcohol dehydrogenase from higher plants involved in lignin biosynthesis, *Phytochemistry*, 13, 2427, 1974.
64. **Kutsuki, H., Shimada, M., and Higuchi, T.,** Regulatory role of cinnamyl alcohol dehydrogenase in the formation of guaiacyl and syringyl lignins, *Phytochemistry*, 21, 19, 1982.
65. **Grand, C., Sarni, F., and Lamb, C. J.,** Rapid induction by fungal elicitor synthesis of cinnamyl-alcohol dehydrogenase, a specific enzyme of lignin synthesis, *Eur. J. Biochem.*, 169, 73, 1987.
66. **Walter, M. H., Grima-Pettenati, J., Grand, C., Boudet, A. M., and Lamb, C. J.,** Cinnamyl-alcohol dehydrogenase, a molecular marker specific for lignin synthesis: cDNA cloning and mRNA induction by fungal elicitor, *Proc. Natl. Acad. Sci. U.S.A.*, 85, 5546, 1988.
67. **Walter, M. H., Grima-Pettenati, J., Grand, C., Boudet, A. M., and Lamb, C. J.,** Extensive sequence similarity of the CAD4 (cinnamyl-alcohol dehydrogenase) to a maize malic enzyme, *Plant Mol. Biol.*, 15, 525, 1990.
68. **Van Doorsselaere, J., Villarroel, R., Van Montagu, M., and Inze, D.,** Nucleotide sequence of a cDNA encoding malic enzyme from poplar, *Plant Physiol.*, 96, 1385, 1991.
69. **O'Malley, D., Porter, S., and Sederoff, R. R.,** Purification characterization and cloning of cinnamyl alcohol dehydrogenase in loblolly pine (*Pinus taeda* L.), *Plant Physiol.*, 98, 1364, 1992.
70. **He, L., Davin, L. B., Eberhardt, T. L., Sederoff, R. R., and Lewis, N. G.,** Cinnamyl alcohol dehydrogenase in the stem and cell-suspension culture of loblolly pine *(Pinus taeda)*, submitted.
71. **Halpin, C., Knight, M. E., Grima-Pettanati, J., Goffner, D., Boudet, A., and Schuch, W.,** Purification and characterization of cinnamyl alcohol dehydrogenase from tobacco stems, *Plant Physiol.*, 98, 12, 1992.
72. **Higuchi, T.,** Lignin biochemistry: biosynthesis and biodegradation, *Wood Sci. Technol.*, 24, 33, 1990.
73. **Sederoff, R. R. and Chang, H.-M.,** Lignin biosynthesis, in *Wood Structure and Composition*, Lewin, M. and Goldstein, I., Eds., Marcel Dekker, New York, 1992, 263.
74. **Harmatha, J., Lübke, H., Rybard, I., and Mähdalik, M.,** *Cis*-coniferyl alcohol and its glucoside from the bark of beech (*Fagus silvatica* L.), *Collect. Czech. Chem. Commun.*, 43, 774, 1977.
75. **Morelli, E., Rej, R. N., Lewis, N. G., Just, G., and Towers, G. H. N.,** *Cis*-monolignols in *Fagus grandifolia* and their possible involvement in lignification, *Phytochemistry*, 25, 1701, 1986.
76. **Lewis, N. G., Dubelsten, P., Eberhardt, T. L., Yamamoto, E., and Towers, G. H. N.,** The *E/Z* isomerization step in the biosynthesis of *Z*-coniferyl alcohol in *Fagus grandifolia*, *Phytochemistry*, 26, 2729, 1987.

77. **Lewis, N. G., Inciong, Ma. E. J., Ohashi, H., Towers, G. H. N., and Yamamoto, E.,** Exclusive accumulation of Z-isomers of monolignols and their glucosides in bark of *Fagus grandifolia, Phytochemistry,* 27, 2119, 1988.
78. **Lewis, N. G., Inciong, Ma. E. J., Dhara, K. P., and Yamamoto, E.,** High-performance liquid chromatographic separation of *E*- and *Z*-monolignols and their glucosides, *J. Chromatogr.*, 479, 345, 1989.
79. **Higuchi, T., Ito, Y., Shimada, M., and Kawamura, J.,** Chemical properties of milled wood lignin of grasses, *Phytochemistry,* 6, 1551, 1967.
80. **Scalbert, A., Monties, B., Lallemand, J.-Y., Guittet, E., and Rolando, C.,** Ether linkage between phenolic acids in lignin fraction from wheat straw, *Phytochemistry,* 24, 1359, 1985.
81. **Terashima, N. and Fukushima, K.,** Biogenesis and structure of macromolecular lignin in the cell wall of tree xylems as studied by microautoradiography, in *Plant Cell Wall Polymers, Biogenesis and Biodegradation,* Lewis, N. G. and Paice, M. G., Eds., *ACS Symp. Ser.*, 399, 160, 1989.
82. **Yamamoto, E., Bokelman, G. H., and Lewis, N. G.,** Phenylpropanoid metabolism in cell walls: an overview, in *Plant Cell Wall Polymers: Biogenesis and Biodegradation,* Lewis, N. G. and Paice, M. G., Eds., *ACS Symp. Ser.*, 399, 68, 1989.
83. **Yamamoto, E., Inciong, Ma. E. J., Davin, L. B., and Lewis, N. G.,** Formation of *cis*-coniferin in cell-free extracts of *Fagus grandifolia* Ehrh bark, *Plant Physiol.*, 94, 209, 1990.
84. **Davin, L. B. and Lewis, N. G.,** Stereoselectivity in polyphenol biosynthesis, in *Plant Polyphenols: Synthesis, Properties and Significance,* Hemingway, R. W. and Laks, P. E., Eds., Plenum Press, New York, 1992, 73.
85. **Lagrimini, L. M., Bradford, S., and Rothstein, S.,** Peroxidase-induced wilting in transgenic tobacco plants, *Plant Cell,* 2, 7, 1990.
86. **Rothstein, S. J. and Lagrimini, L. M.,** Silencing gene expression in plants, in *Oxford Surveys of Plant Molecular and Cell Biology,* Miflin, B. J. and Miflin, H. F., Eds., Oxford University Press, New York, 1989, 221.
87. **Lagrimini, L. M.,** Altered phenotypes in plants transformed with chimeric tobacco peroxidase genes, in *Molecular and Physiological Aspects of Plant Peroxidases.* II, Penel, C., Gaspar, T., and Lobarzewiki, J., Eds., in press.
88. **Sterjiades, R., Dean, J. F. D., and Eriksson, K. E. L.,** Laccase from sycamore maple (*Acer pseudoplatanus*) polymerizes monolignols, *Plant Physiol.*, in press.
89. **Freudenberg, K., Harkin, J., Reichert, M., and Fukuzumi, T.,** Die an der verholzung beteiligten enzyme. Die dehydrierung des sinapinalkohols, *Chem. Ber.*, 91, 581, 1958.
90. **Higuchi, T. and Ito, Y.,** Dehydrogenation productions of coniferyl alcohol formed by the action of mushroom phenol oxidase, rhus laccase and radish peroxidase, *J. Biochem. (Tokyo),* 45, 575, 1958.
91. **Higuchi, T.,** Further studies on phenol oxidase related to the lignin biosynthesis, *J. Biochem. (Tokyo),* 45, 515, 1958.
92. **Nakamura, W.,** Studies on the biosynthesis of lignin. I. Disproof against the catalytic activity of laccase in the oxidation of coniferyl alcohol, *J. Biochem. (Tokyo),* 62, 54, 1967.
93. **Savidge, R. A. and Randeniya, P. U.,** Evidence for coniferyl-alcohol oxidase promotion of lignification in developing xylem of conifers, *Biochem. Soc. Trans.*, 20, 2295, 1992.
94. **Savidge, R. A. and Udagama-Randeniya, P.,** Cell-wall bound coniferyl alcohol oxidase associated with lignification in conifers, *Phytochemistry,* 1992.
95. **Gross, G. G., Janse, C., and Elstner, E. F.,** Involvement of malate, monophenols, and the superoxide radical in hydrogen peroxide formation by isolated cell walls from horseradish (*Armoracia lapathifolia* Gilib), *Planta,* 271, 1977.
96. **Elstner, E. F. and Heupel, A.,** Formation of hydrogen peroxide by isolated cell walls from horseradish (*Armoracia lapathifolia* Gilib), *Planta,* 130, 175, 1976.
97. **Gross, G. G.,** Cell-wall bound malate dehydrogenase from horseradish, *Phytochemistry,* 16, 319, 1977.

98. **Gross, G. G. and Janse, C.,** Formation of NADH and hydrogen peroxide by cell wall associated enzymes from *Forsythia* xylem, *Z. Pflanzenphysiol.*, 845, 447, 1977.
99. **Lewis, N. G., Newman, J., Rej, R. N., Just, G., and Ripmeister, J.,** Determination of bonding patterns of ^{13}C specifically enriched DHP lignin in solution and solid state, *Macromolecules*, 20, 1752, 1987.
100. **Lewis, N. G., Yamamoto, E., Wooten, J. B., Just, G., Ohashi, H., and Towers, G. H. N.,** Monitoring biosynthesis of wheat cell-wall phenylpropanoids *in situ*, *Science*, 237, 1344, 1987.
101. **Lewis, N. G., Razal, R. A., Dhara, K. P., Yamamoto, E., Bokelman, G. H., and Wooten, J. B.,** Incorporation of [2-^{13}C]ferulic acid, a lignin precursor, into *Leucaena leucocephala* and its analysis by solid-state ^{13}C-NMR, *J. Chem. Soc. Chem. Commun.*, 1626, 1988.
102. **Lewis, N. G.,** Lignin biosynthesis, biodegradation and utilization, *Bull. Liaison Groupe Polyphenols*, 14, 398, 1988.
103. **Lewis, N. G., Razal, R. A., Yamamoto, E., Bokelman, G. H., and Wooten, J. B.,** ^{13}C-specific labelling of lignin in intact plants, in *Plant Cell Wall Polymers: Biogenesis and Biodegradation*, Lewis, N. G. and Paice, M. G., Eds., *ACS Symp. Ser.*, 399, 169, 1989.
104. **Whiting, D. A. I.,** Lignans and neolignans, *Nat. Prod. Rep.*, 2, 191, 1985.
105. **Rao, C. B. S.,** *Chemistry of Lignans*, Andrha University Press, Waltair, India, 1978, 377 pp.
106. **Davin, L. B., Bedgar, D. L., Katayama, T., and Lewis, N. G.,** On the stereoselective synthesis of (+)-pinoresinol in *Forsythia suspensa* from its achiral precursor, coniferyl alcohol, *Phytochemistry*, 31, 3869, 1992.
107. **Katayama, T., Davin, L. B., and Lewis, N. G.,** An extraordinary accumulation of (−) pinoresinol in cell-free extracts of *Forsythia* intermedia: evidence for enantiospecific reduction of (+)-pinoresinol, *Phytochemistry*, 31, 3875, 1992.
108. **Umezawa, T., Davin, L. B., Kingston, D. G. I., Yamamoto, E., and Lewis, N. G.,** Lignan biosynthesis in *Forsythia* sp., *J. C. S. Chem. Commun.*, 1405, 1990.
109. **Umezawa, T., Davin, L. B., and Lewis, N. G.,** Formation of the lignan, (−)-secoisolariciresinol by cell-free extracts of *Forsythia intermedia*, *Biochem. Biophys. Res. Commun.*, 171, 1008, 1990.
110. **Umezawa, T., Davin, L. B., and Lewis, N. G.,** Formation of lignans, (−)-secoisolariciresinol and (−)-matairesinol with *Forsythia intermedia* cell-free extracts, *J. Biol. Chem.*, 266, 10210, 1991.
111. **Hartley, R. D. and Ford, C. W.,** Phenolic constituents of plant cell walls and wall biodegradability, in *Plant Cell Wall Polymers: Biosynthesis and Biodegradation*, Lewis, N. G. and Paice, M. G., Eds., *ACS Symp. Ser.*, 399, 137, 1989.
112. **Stafford, H. A. and Brown, M. A.,** Oxidative dimerization of ferulic acid by extracts of sorghum, *Phytochemistry*, 15, 465, 1976.
113. **Brown, S. A.,** Biochemistry of plant coumarins, in *The Shikimic Acid Pathway*, Conn, E. E., Ed., *Rec. Adv. Phytochem.*, 20, 287, 1985.
114. **Gestetner, B. and Conn, E. E.,** 2-Hydroxylation of *trans*-cinnamic acid by chloroplasts from *Melilotus alba*, *Arch. Biochem. Biophys.*, 163, 617, 1974.
115. **Ranjeva, R., Alibert, G., and Boudet, A. M.,** Metabolisme des composés phenolicues chez le petunia. V. Utilization de la phenylalanine par des chloroplastes, isolés, *Plant Sci. Lett.*, 10, 225, 1977.
116. **Conn, E. E.,** Compartmentalization of secondary compounds, in *Membranes and Compartmentation in the Regulation of Plant Functions*, Boudet, A. M., Alibert, G., and Lea, P. J., Eds., Oxford University Press, Oxford, *Annu. Proc. Phytochem. Soc. Eur.*, 24, 1, 1984.
117. **Haskins, F. A., Williams, L. G., and Gorz, H. J.,** Light-induced *trans* to *cis* conversion of β-D-glucosyl-*o*-hydroxycinnamic acid in *Melilotus alba* leaves, *Plant Physiol.*, 39, 777, 1964.
118. **Brown, S. A.,** Biosynthesis of 6,7-dihydroxy coumarin in *Cichorium intybus*, *Can. J. Biochem. Cell Biol.*, 63, 292, 1985.
119. **Bernards, M. A. and Lewis, N. G.,** Alkyl ferulates in wound healing potato tubers, *Phytochemistry*, 31, 3409, 1992.

120. **Guillemonat, A. and Traynard, J.-C.**, Sur la constitution chimique du liége. IV. Mémoire: structure de la phellochryseine. Memoires présentés á la société, *Chimique*, 142, 1962.
121. **Adamovics, J. A., Johnson, G., and Stermitz, F. R.**, Ferulates from cork layers of *Solanum tuberosum* and *Pseudotsuga menziesii, Phytochemistry*, 16, 1089, 1977.
122. **Laver, M. L. and Fang, H. L.**, Ferulic acid esters from bark of *Pseudotsuga menziesii, J. Agric. Food Chem.*, 37, 114, 1989.
123. **Borchert, R.**, Time course and spatial distribution of phenylalanine ammonia-lyase and peroxidase activity in wounded potato tuber tissue, *Plant Physiol.*, 62, 789, 1978.
124. **Espelie, K. E. and Kolattukudy, P. E.**, Purification and characterization of an abscisic acid-inducible anionic peroxidase associated with suberization in potato *(Solanum tuberosum), Arch. Biochem. Biophys.*, 240(2), 539, 1985.
125. **Roberts, E., Kutchan, T., and Kolattukudy, P. E.**, Cloning and sequence of cDNA for a highly anionic peroxidase from potato and the induction of its mRNA in suberizing potato tubers and tomato fruits, *Plant Mol. Biol.*, 11(1), 15, 1988.
126. **Roberts, E. and Kolattukudy, P. E.**, Molecular cloning, nucleotide sequence and abscisic acid induction of a suberization-associated highly anionic peroxidase, *Mol. Gen. Genet.*, 217, 223, 1989.
127. **Kolattukudy, P. E. and Espelie, K.-E.**, Biosynthesis of cutin, suberin and associated waxes, in *Biosynthesis and Biodegradation of Wood Components*, Higuchi, T., Ed., Academic Press, New York, 1985, 161.
128. **Kolattukudy, P. E.**, Biochemistry and function of cutin and suberin, *Can. J. Bot.*, 62, 2918, 1984.
129. **Matzke, M. and Riederer, M. A.**, Comparative study into the chemical composition of cutins and suberins from *Picea abies* (L) Karst, *Quercus robur* L. and *Fagus sylvatica* L., *Planta*, 185, 233, 1991.
130. **Hergert, H. L.**, Chemical composition of cork from white fir bark, *For. Prod. J.*, 335, 1958.
131. **Kolattukudy, P. E., Kronman, K., and Poulouse, A. J.**, Determination of structure and composition of suberin from the roots of carrot, parsnip, rutabaga, turnip, red beet and sweet potato by combined gas-liquid chromatography and mass spectrometry, *Plant Physiol.*, 55, 567, 1975.
132. **Holloway, P. J.**, The composition of suberin from the corks of *Quercus suber* L. and *Betula pendula* Roth, *Chem. Phys. Lipids*, 9, 158, 1972.
133. **Litvay, J. D. and Krahmer, R. G.**, Wall-layering in Douglas fir cork cells, *Wood Sci.*, 9, 167, 1977.
134. **Garbow, J. R., Ferrantello, L. M., and Stark, R. E.**, ^{13}C Nuclear magnetic resonance study of suberized potato cell wall, *Plant Physiol.*, 90, 783, 1989.
135. **Stark, R. E., Zlotnik-Mazori, T., Ferrantello, L. M., and Garbow, J. R.**, Molecular structure and dynamics of intact plant polyesters and solid-state NMR studies, in *Plant Cell Wall Polymers: Biogenesis and Biodegradation*, Lewis, N. G. and Paice, M. G., Eds., *ACS Symp. Ser.*, 399, 214, 1989.
136. **Zimmerman, W., Nimz, H., and Seemüller, E.**, ^{1}H and ^{13}C NMR spectroscopy of extracts from corks of *Rubus idaeus, Solanum tuberosum*, and *Quercus suber, Holzforschung*, 39, 45, 1985.
137. **Cooke, G.**, *Cork and Cork Products*, Crown Cork and Seal Co., 1942, 30.
138. **Jensen, W. and Ostman, R.**, Paper and timber, *Finland*, 36, 427, 1954.
139. **Sharkov, W., Kalina, V., and Sobetskii, S.**, *Lesokhim. Prom. No.* 5, 8, 1939; *Chem. Ab.*, 34, 5273, 1940.
140. **Jensen, W., Ihalo, P., and Varska, K.**, *Pap. Puu.*, 39, 237, 1957.
141. **Jensen, W., Fremer, K.-E., Sierildä, P., and Wartiovaara, V.**, The chemistry of bark, in *The Chemistry of Wood*, Browning, B. L., Ed., Interscience, New York, 1963, 587.
142. **Kolattukudy, P. E. and Dean, B. B.**, Structure and gas chromatographic measurement, and function of suberin synthesized by potato tuber tissue slices, *Plant Physiol.*, 54, 116, 1974.

143. **Soliday, C. L., Kolattukudy, P. E., and Davis, R. W.,** Chemical and ultrastructural evidence that waxes associated with the suberin polymer constitute the major diffusion barrier to water vapour in potato tuber (*Solanum tuberosum* L.), *Planta,* 146, 607, 1979.
144. **Ribas, I.,** *Ion,* 2(6), 25, 1942.
145. **Swan, E. P.,** *TAPPI,* 51, 301, 1968.
146. **Riley, R. G. and Kolattukudy, P. E.,** Evidence for covalently attached *p*-coumaric acid and ferulic acid in cutins and suberins, *Plant Physiol.,* 56, 650, 1975.
147. **Hahlbrock, K. and Scheel, D.,** Physiology and molecular biology of phenylpropanoid metabolism, *Annu. Rev. Plant Physiol. Plant Mol. Biol.,* 40, 347, 1989.
148. **Cottle, W. and Kolattukudy, P. E.,** Biosynthesis, deposition and partial characterization of potato suberin phenolics, *Plant Physiol.,* 56, 650, 1982.
149. **Borg-Olivier, O. and Monties, B.,** Caractérisation des lignines, acides phénoliques et tyramine dans les tissus subérisés du périderme naturel et du périderme de blessure de tubercule de pomme de terre, *C.R. Acad. Sci. Paris, Ser. III,* 141, 1989.

INDEX

E

F

G

H

I

J

L

M

N

O

P